FOOD SCIENCE AND TECHNOLOGY

APPLICATIONS OF ULTRASOUND IN THE BEVERAGE INDUSTRY

FOOD SCIENCE AND TECHNOLOGY

Additional books in this series can be found on Nova's website
under the Series tab.

Additional e-books in this series can be found on Nova's website
under the eBooks tab.

FOOD SCIENCE AND TECHNOLOGY

APPLICATIONS OF ULTRASOUND IN THE BEVERAGE INDUSTRY

JUAN FRANCISCO GARCÍA MARTÍN, PHD
EDITOR

nova publishers
New York

Additional color graphics may be available in the e-book version of this book.

Library of Congress Cataloging-in-Publication Data
Names: García Martín, Juan Francisco, editor.
Title: Applications of ultrasound in the beverage industry / editor,
Juan Francisco García Martín, Department of Chemical Engineering,
University of Málaga, Spain.
Description: Hauppauge, New York : Nova Science Publishers, Inc., [2016] |
Series: Food science and technology | Includes index.
Identifiers: LCCN 2016012396 | ISBN 9781634850698 (hardcover)
Subjects: LCSH: Ultrasonic waves--Industrial applications. | Beverages. |
Beverage industry. | Food industry and trade. | Sonochemistry. |
Food--Preservation.
Classification: LCC TP372.4 .A67 2016 | DDC 663--dc23 LC record available at
http://lccn.loc.gov/2016012396

Published by Nova Science Publishers, Inc. † New York

CONTENTS

PREFACE

Ultrasound is regarded as an emerging and promising technology in industrial food processing. In recent years, several applications of ultrasound to the production of beverages, mainly wines and juices, have been assayed at laboratory scale. To be specific, ultrasound has been used for enhancing heat transfer, microbial contamination detection, reducing membrane fouling in beverage clarification, microorganism inactivation, equipment cleaning, process monitoring, assisted extraction and accelerating reactions within beverages. This book presents an overview of the applications of ultrasound in beverage industries. The most recent developments are discussed and the future prospects for research in this field are explored.

Ultrasound is a non-thermal method in food processing that has the advantage of preserving fruit juices without causing the common side-effects associated with conventional heat treatments. It is a simple and effective method for retaining the original characteristics of fruit juices, which is an advantage in relation to other methods such as pasteurization. Therefore, ultrasound can be used as pre-treatment method (drying, freezing) in juice production. Besides, the physical and chemical effects of cavitation are considered to be responsible for the ultrasound-induced inactivation of enzymes, which depends on the type of enzyme, the medium in which the enzyme is suspended, and the processing conditions. In *Chapter 1*, Režek Jambrak and Vukušić summarize the main effects of ultrasound on fruit, vegetable and other formulated beverages, and its potential applications in enzyme inactivation, extraction of bioactive compounds (phenols, carotenoids, anthocyanins), processing of residues from fruit and vegetable juice processing, and as a pre-treatment method (prior spray-drying, freezing, lyophilisation etc.). These authors briefly sum up the consequences of applying ultrasound to fruit juices in relation to their aroma and sensory profile.

The effects of ultrasound on microorganisms are both physical (cavitation, mechanical effects, micromechanical shocks) and chemical (due to formation of free radicals formed by the decomposition of water inside the oscillating bubbles). In order to enhance its effects, ultrasound can be used together with pressure treatment (manosonication), heat treatment (thermosonication) or both (manothermosonication). Many researchers have demonstrated excellent effects in inactivation of microorganisms in the beverage industry. In *Chapter 2,* Režek Jambrak et al. make an overview of the potential of ultrasound in inactivation of microorganisms in fruit juices by showing the main results from researches that can be found to date in literature.

The beverage industry produces million tons of byproducts per year, which represents a major disposal problem. As aforementioned in Chapter 1, these byproducts can be used as sources of various bioactive compounds, including flavonoids, carotenoids, anthocyanins, vitamins, polysaccharides and many other phytochemicals. These bioactive compounds can come from fruits or can be added into juices to produce functional fruit beverages. Because of their health-beneficial effects, the bioactive compounds recovered from fruit beverage byproducts can be potentially used in food and pharmaceutical industries. In *Chapter 3,* Yang et al. illustrate the use of ultrasound, as an innovating and green extraction technology, to recover bioactive compounds from fruit beverage derived byproducts. These authors expose the various factors that can affect the efficiency of ultrasound during extraction. These factors can be divided into ultrasonic (ultrasound energy, ultrasound duration, ultrasound frequency and ultrasound treatment mode) and non-ultrasonic factors (solvent type, particle size, extraction temperature, and solvent-to-solid ratio). Finally, Yang et al. summarize some mathematical models that can be used to simulate the ultrasound-assisted extraction process and therefore to control the extraction process and to scale up the ultrasound-assisted extractor.

Food spoilage is a major cause of economic loss in the beverage industry and can be attributed to the growth of microorganisms or the activity of the enzymes presented in the fruits. Although thermal processing is generally used for controlling the safety standards, this approach may lead to quality reduction of the processed juice as well as considerable processing costs. As stated in Chapter 2, ultrasound has a great inactivation effect on microorganisms in fruit juices. *Chapter 4* covers different aspects related to the application of ultrasound for sterilization of fruit and vegetable juices. In this chapter, Gogate and Khandpur illustrate the mechanism of microbial disinfection and the fundamentals of ultrasound-based reactor designs, and provide guidelines for important operating parameters. These authors also provide a comparison of ultrasound efficacy with those of conventional approaches. Lastly, Gogate and Khandpur make an overview of available literature for deactivation using ultrasound and ultraviolet radiation and highlight the synergistic results they obtained in their research on the combination of ultrasound with ultraviolet radiation and orange peel extract.

In recent years, some physical technologies have been assayed to accelerate reactions within wine in order to achieve a high-quality wine resembling a wine that has been subjected to many years of natural ageing. Among these physical ageing techniques, ultrasound is regarded as the most promising. However, from the information available in literature it can be concluded that the application of ultrasound to wine is not enough to accelerate the wine ageing process. *Chapter 5* and *Chapter 6* illustrate the effects of ultrasound on wine from opposed point of views. In Chapter 5, my collaborators and I summarize the chemical effects that we have found in our recent research on the application of ultrasound to red wine. Besides, we also review other potential applications of ultrasound in the winemaking industry. In Chapter 6, Nolan, as a highly respected wine professional, offers his vision about what consumers can expect from artificially aged wines.

Finally, and closely related to Chapters 5 and 6, Ferraretto and Celotti present their last research paper in *Chapter 7*. These authors studied the extraction of phenolic compounds from grapes subjected to ultrasound. The disruption of the cell wall due to pressure alternance and cavitation was responsible for phenolic compound release. Afterwards, Ferraretto and Celotti assessed the application of ultrasound to red wine in order to accelerate the ageing process. From data obtained on wines after ultrasound treatment, these authors suggest the use of the ultrasound to promote the polymerization of the phenolic compounds as the wine matures and therefore accelerate the wine ageing process. All in all, the application of ultrasound in winemaking represents a possibility to optimize fermentation technology and, therefore, a better management of red grape vinification. On the other hand, the application of ultrasound can be considered as a continuous pre-treatment of crushed red grapes before loading the vinification tank, due to the very short required ultrasound exposure times.

In: Applications of Ultrasound in the Beverage… ISBN: 978-1-63485-069-8
Editor: Juan Francisco García Martín

Chapter 1

State of the Art of the Use of Ultrasound in the Beverage Industry (I): Effects on Beverages

Anet Režek Jambrak*[*] *and Tomislava Vukušić
Faculty of Food Technology and Biotechnology,
Zagreb, Croatia

Abstract

Power ultrasound is non-thermal technique that has been extensively used in last ten years in food processing. Power ultrasound is used in processing of juices, as well in formulation of beverages. The main usage of power ultrasound in beverages industry is in inactivation of microorganisms and achievement of 5-log reduction (as requested by the U.S. Food and Drug Administration). Also, power ultrasound could be used in inactivation of enzymes, extraction of bioactive compounds (phenols, carotenoids, anthocyanins), processing of residues (waste) from fruit and vegetable juice processing, and as a pre-treatment method (prior spray-drying, freezing, lyophilisation etc.). When applying ultrasound it is important to observe changes in aroma profile and sensory properties of beverages, to observe radical formation, and to prevent deterioration of extractive compounds. Power ultrasound could be used in homogenization and preservation of beverages that are prepared

[*] Corresponding Author Email: arezek@pbf.hr.

according to specific formulation (whey protein beverages, whey based beverages, dairy, smoothies, vegetable juices).

Keywords: power ultrasound, fruit, juice, extraction, properties

INTRODUCTION

Ultrasonication is a non-thermal method in food processing that has the advantage of preserving fruit juices without causing the common side-effects associated with conventional heat treatments (Režek Jambrak and Herceg, 2014).

High-intensity ultrasound is used in many food applications, such as emulsifying, sterilizing, extracting, degassing, filtrating, drying, and enhancing oxidation (Režek Jambrak and Herceg, 2014). High intensity ultrasound generated by periodic mechanical motions of a probe, transfers ultrasonic energy into a fluid medium and triggers extremely high alterations in pressure leading to the formation of small rapidly growing bubbles (cavities) (Mason, 1996), which expand during the negative pressure excursion, and implode violently during the positive excursion generating high temperatures, high pressures and shear forces at the probe tip. This phenomenon is known as cavitation. During implosion, very high temperatures (approximately 5500 K) and pressures (approximately 50 MPa) are reached inside these bubbles (Mason, 1996, Šimunek et al., 2013) that is consequently causing several reactions around imploding bubble. There are four types of cavitation based on the mode of generation: acoustic, optic, particle and hydrodynamic. In processing application, like in food processing industry, only acoustic and hydrodynamic cavitation are found to be efficient (Gogate and Kabadi, 2009), because of producing chemical or physical changes in the treated material.

Applications of ultrasound in processing of fruit juices and the effects of sonication on fruit juices were studied (Režek Jambrak and Herceg, 2014). Sonication is considered one of the potential non-thermal processing techniques for the processing of fruit juices. It is simple and effective method of retaining the original characteristics of fruit juices, which is an advantage in relation to pasteurization (Šimunek et al., 2013). Sonication technology can improve the process by reduced processing time, higher throughput and lower energy consumption. If ultrasound is intended to be used in any practical application, it would most likely have to be used in conjunction with pressure treatment (manosonication), heat treatment (thermosonication) or both (manothermosonication). The effect of ultrasound was mainly attributed to physical (cavitation, mechanical effects, micro-mechanical shocks) and/or chemical effects due to formation of free radicals (H^+ and OH^- due to sonochemical reaction) formed by the decomposition of water inside the oscillating bubbles.

FRUIT, VEGETABLE AND OTHER FORMULATED BEVERAGES

Fruit and vegetable juices are the unfermented but fermentable liquids obtained from the edible part of sound, appropriately mature and fresh fruits and vegetables and prepared by suitable processes, which maintain the essential physical, chemical, organoleptic and nutritional characteristics of the juices. As consumers are becoming more health conscious, demands for natural and fresh-like foods have increased interest in using non-thermal technologies in food processing to produce foods with minimal damage to nutritional and sensory properties. Sonication is regarded to be useful for minimal processing and could affect a process with reduced processing time, higher throughput and lower energy consumption as well as enhance extraction yield and nutritional quality of juices (Cheng et al., 2007; Mason, 1996; Zenker et al., 2003).

It has been reported to be effective against foodborne pathogens found in orange juice (Valero et al., 2007), guava juice (Cheng et al., 2007). Ultrasound processing of juices is reported to have minimal effect on the degradation of key quality parameters such as color and ascorbic acid in orange juice during storage at 10°C (Tiwari et al., 2009).

Although sonication treatments induced changes in all the color values of grapefruit juice, these changes were not easily seen by the naked eyes, therefore, it is suggested that the sonication technique might be employed for the processing of grapefruit juice (Aadil et al., 2013). There are several studies reporting increase of vitamin C in sonicated citrus juices which is highly beneficial to the human health (Aadil et al., 2013; Bhat et al., 2011), but also reporting degradation of ascorbic acid in tomato juice (Adekunte et al., 2010), as a result of oxidation reactions, promoted by the interaction with free radicals formed during sonication (Hart and Henglein, 1985).

This positive effect of ultrasound is assumed to be due to the effective removal of occluded oxygen from the juice (Knorr et al., 2004). The effect of ultrasound on anthocyanins was studied for strawberry juice by Tiwari et al. (2008). The Tiwari's group reported how ultrasound processing of juices has a minimal effect on the degradation of key quality parameters such as color and ascorbic acid and anthocyanin content in strawberry juice and anthocyanin content of blackberry juice. They reported a slight increase (1-2%) in the pelargonidin-3-glucoside content of the juice at lower amplitude levels and treatment times, which may be due to the extraction of bound anthocyanins from the suspended pulp.

Ultrasonic is one of the common methods in enhancing mass transfer (Lou et al., 2011). Ultrasound assisted extraction of bioactive compounds and anthocyanins were recently reviewed by Vilkhu et al. (2008) where is stated how ultrasound assisted extraction improves the extraction yield of bioactive compounds by between 6% and 35% (Vilkhu et al., 2008). Sonication is especially beneficial in processing of soy beverages with solid parts from blended juices where extraction of isoflavons is simpler and done in reproducible low cost, low toxicity and environmental way (Rostagno et al., 2003, 2007).

EFFECT OF ULTRASOUND ON ENZYMES

High-power ultrasound is a versatile technology which can potentially be used in many food processing applications. Ultrasound treatment alone does not efficiently cause enzyme inactivation. However, combined with mild heat with or without elevated pressure ($p \leq 500$ kPa), ultrasound can effectively inactivate enzymes. The application of ultrasound has been shown to enhance the rate of inactivation of quality degrading enzymes including pectin methyl esterase (PME), polygalacturonase (PG), peroxidase (POD), polyphenol oxidase (PPO), and lipoxygenase (LOX) at mild temperature (40-50°C) by up to 400 times. Ultrasound also enables the inactivation of relatively heat-resistant enzymes such as tomato PG1 and thermostable orange PME at mild temperature conditions.

The extent to which ultrasound enhances the inactivation rate depends on the type of enzyme, the medium in which the enzyme is suspended, and the processing condition including frequency, ultrasonic intensity, temperature, and pressure (Shiferawterefe et al., 2015).

The physical and chemical effects of cavitation are considered to be responsible for the ultrasound-induced inactivation of enzymes, although the dominant mechanism depends on the structure of the enzyme. Combined treatments with mild heat and mild pressure have been shown to be the best approach in this regard. Experimental investigations have shown that free radical mediated mechanisms play a significant role in the ultrasonic inactivation of many enzymes.

In the study by Safari et al. (2013) the potential role of hydrogen peroxide (H_2O_2) as a signaling molecule in regulation of activity of stress-related enzymes, and expression of catalase (CAT) and phenylalanine ammonialyase (PAL) genes was investigated. Suspension-cultured *Corylus avellana* L. cells were agitated by an ultrasonic device at 29 kHz with the power of 4 mW/cm^2, for 8–40 min. The activities of CAT, superoxide dismutase (SOD), and ascorbate peroxidase (APX) of treated cells increased by 4, 1.7 and 7 times of the control ones, respectively. Induction of increase in the expression of CAT gene started 24 h after the treatment with ultrasound. Significant increase was also observed in the expression of PAL gene, 6 h after exposure to ultrasound, which resulted in turn to increase of total contents of soluble phenolics after 24 h of treatment. Exposure to ultrasound up to 20 min had no adverse effects on cell viability. It is likely that level of increased H_2O_2 was not deteriorative for hazel cells, but rather triggered antioxidant system and provided hazel cells a sustainable growth after ultrasound treatment. The results suggest that the activation of PAL may be regulated by the H_2O_2 accumulation during the elicitation by ultrasound. Exposure to low-energy ultrasound can increase the level of H_2O_2 so that not being sufficient to promote phytotoxic effects, but sufficient to promote expression of defense genes (Safari et al., 2013).

In the study by Subhedar et al. (2015) the effect of ultrasound on enzymatic hydrolysis of newspapers, a potential feedstock for bioethanol production, was investigated. In case of ultrasound assisted-enzymatic hydrolysis approach, optimum conditions obtained were substrate loading of 3% (w/v), enzyme loading of 0.8% (w/v), sonication power of 60 W, duty cycle of 70%, and hydrolysis time of 6.5 h, and the reducing sugars yield obtained under these conditions was 27.6 g/L. Approximately 2.4 times increase in the release of reducing sugars concentration was obtained by the ultrasound-assisted enzymatic hydrolysis approach. Results indicate that there is a synergistic effect obtained from the combination of ultrasound and enzymes which lowers the diffusion-limiting barrier to enzyme/substrate binding and results in an increase in reaction rate.

Extraction, Effect of Ultrasound on Bioactive Compounds (Phenols, Carotenoids, Anthocyanins) and Antioxidant Activity

Ultrasound processing of juices is reported to have a minimal effect on the degradation of key quality parameters such as color and anthocyanin content in strawberry and blackberry juices. The effect of ultrasound on anthocyanins was studied for strawberry juice by Tiwari et al. (2008). They reported a slight increase (1–2%) in the pelargonidin-3-glucoside (P3G) content of the juice at lower amplitude levels and treatment times, which may be due to the extraction of bound anthocyanins from the suspended pulp. However, in a study by Tiwari et al. (2009), the anthocyanin content of the juice was degraded when higher amplitude levels were employed, but the maximum observed degradation was <5%.

In the study by Jambrak et al. (2013) (unpublished data) the effect of power ultrasound on antioxidant capacity, color, total phenolic compounds and degree of non-enzymatic browning (NEB) of blueberry and cranberry nectar was investigated. Samples were treated with power ultrasound at 100%, 75% and 50% amplitude with treatment time 3, 6 and 9 minutes. For planning and optimizing treatment parameters, full factorial design and response surface method was used. Antioxidant capacity was determined by DPPH and ABTS methods. Their comparison shows a positive correlation between the values of antioxidant capacity. For ABTS method it was observed that treatment time has statistically significant impact on blueberry nectar antioxidant capacity. Cranberry nectar color was the darkest and with highest levels of red and yellow in untreated sample, while in blueberry nectar the sample treated with ultrasound at 100% amplitude for 6 minutes had similar features. The degree of non-enzymatic browning was determined spectrophotometric and it was the highest for samples treated with ultrasound at 100% amplitude for 9 minutes, and in these similar conditions also total phenolic compounds concentration increased.

In the study by Long et al. (2011) an aqueous enzymatic process assisted by ultrasound extraction (AEP-UE) was applied to the extraction of oil from flaxseed (*Linum usitatissimum* L.). The AEP-UE-derived oil had a 1.5% higher content of unsaturated fatty acids than the solvent extraction-derived oil. The aqueous enzymatic ultrasound extraction oil showed higher DPPH radical-scavenging activity. AEP-UE derived oil exhibited higher antioxidant activity.

Khandpur and Gogate (2015) studied the effects of novel approaches based on the use of ultrasound on the nutritional quality of different fruit and vegetable juices (orange, sweet lime, carrot and spinach juices) and its comparison with the conventional thermal pasteurization operated at 80°C for 10 min. The ultrasound sterilization parameters were maintained at ultrasound frequency of 20 kHz and power of 100 W with treatment time as 15 min. The treated juices were analyzed for total phenol content, antioxidant activity, vitamin C, carbohydrates, etc. It has been established that ultrasound processed juice retained most of the nutrient components to higher extent in comparison to all the other techniques used in the work. Combination of ultrasound and ultraviolet irradiations used to achieve an effective decontamination of juices (recommended 5 log reduction of microorganisms) also retained nutrients to a higher level in comparison to the thermal method; however some losses were observed as compared to the use of solely ultrasound which could be attributed to inefficient heat exchange in the combined approach.

Aadil et al. (2013) studied the effect of sonication treatments on some important quality parameters of grapefruit juice such as physico-chemical (pH, acidity and Brix), Hunter color values (L*, a* and b*), cloud value, electrical conductivity, total antioxidant capacity, DPPH (2,2-diphenyl-1-picrylhydrazyl) free radical scavenging activity, ascorbic acid, total phenolics, flavonoids and flavonols. Sonication of grapefruit juice was performed in a bath type sonicator at a frequency of 28 kHz by maintaining a constant temperature of 20ºC. Results showed that there was significant improvement in the cloud value, total antioxidant capacity, DPPH free radical scavenging activity, ascorbic acid, total phenolics, flavonoids and flavonols in all the juice samples sonicated for 30, 60 and 90 min but no changes occurred in the pH, acidity and °Brix value as compared to control. Some differences in all the color values were also observed but overall quality of grapefruit juice was improved, suggesting that sonication technique may successfully be implemented an industrial scale for the processing of grapefruit juice.

Dubrović et al. (2011) studied the influence of high intensity ultrasound and pasteurization on the stability of anthocyanins and their content in strawberry juice. Different ultrasound process parameters for the treatment of juices were compared to the classical thermal treatments. For ultrasound treatments, three parameters were varied according to the statistical experimental design. Central composite design was used to optimize and design experimental parameters: temperature (25, 40 and 55ºC), amplitude (60, 90 and 120 μm) and time (3, 6, and 9 min). It was found that the anthocyanin content after pasteurization (85°C for 2 min) was reduced by 5.3 to 5.8% compared to untreated juices. After treatment with ultrasound (20°C for 3, 6 or 9 min) or thermosonication (40°C for 3, 6 or 9 min and 60°C for 3 or 6 min), the degradation of anthocyanins was generally less intensive and was 0.7–4.4% compared to the untreated juices. Only in the case of ultrasonic treatment at a temperature of 55°C and treatment time of 9 min the total content of anthocyanins, compared to untreated juice, was reduced by 5.8 to 7.1%, and their degradation was greater than that of pasteurized juices. From the results it can be concluded that total anthocyanin content was greater in more than 85% of the selected ultrasound treatments compared to pasteurized juices. Ultrasound treatment can replace pasteurization in terms of preserving total anthocyanin content.

In study by Aadil and Rupasinghe (2012) different apple-carrot juice blends (60:40, 75:25 and 90:10, v/v) were treated with ultrasound and compared to the conventional thermal pasteurization. Juice quality parameters such as color, pH, titratable acid, total soluble solids, antioxidant capacity and beta-carotene did not change significantly during the storage period.

Apple juice was subjected to ultrasound treatment at 30, 60 and 90% in the study by Abid et al. (2014). Amplitude levels were assayed using a probe-type sonicator (20 kHz, 20°C for 3 min) keeping pulse durations of 5 s on/off, and the resulting samples after ultrasound treatment were stored at 4C for 30 days. Effects of sonication and storage on total soluble solids (TSS), pH, titratable acidity, instrumental color (*L**, *a**, *b**), non-enzymatic browning (NEB), cloud value, ascorbic acid and antioxidant capacity were evaluated. As results, TSS, pH and acidity remained unchanged while significant increases in color values, NEB, cloud value, ascorbic acid and antioxidant capacity were observed in all sonicated samples. Storage study showed significant increase in acidity, color values, NEB and decrease in pH, cloud value, ascorbic acid and antioxidant capacity. The results were further verified by principal component analysis. The pH, acidity, NEB, color and cloud values, ascorbic acid and antioxidant capacity were significantly affected during storage period while sonication treatments (amplitude levels of 60 and 90%) improved and preserved cloud value, ascorbic acid content and antioxidant capacity of apple juice. These authors concluded that the technique may potentially be implemented for apple juice processing to preserve its nutritional and sensory quality for consumer's satisfaction (Abid et al., 2014).

Changes in bioactive compounds and shelf-life of carrot juice have been investigated by Martínez-Flores et al. (2015). Carrot juice was thermo-sonicated (24 kHz, 120 μm amplitude) at 50°C, 54°C and 58°C for 10 min (acoustic power 2204.40, 2155.72, 2181.68 mW/mL, respectively). Quality parameters were evaluated after processing and during storage at 4°C. Control and sonicated treatments at 50°C and 54°C had 10, 12 and 14 d of shelf-life, respectively. Samples sonicated at 58°C had the best quality. Thermo-sonicated juice at 58°C retained more than 98% of carotenoids and 100% of ascorbic acid. Phenolic compounds increased in all stored, treated juices.

The stability of lycopene was evaluated after application of high power low frequency ultrasound (Oliveira et al., 2015). The study was carried out on a solution containing pure lycopene to evaluate the direct effect of ultrasound on lycopene and on tomato purée to evaluate the direct and indirect effects of ultrasound application within a food matrix. Power densities ranging from 55 to 5000 W/L and temperatures ranging from 23°C (ambient) to 60°C were evaluated. The experiments on pure lycopene showed that the application of ultrasound did not exert any direct effect over lycopene. However, the retention of lycopene in tomato purée decreased indicating an indirect effect on lycopene stability caused by high concentration of hydrogen peroxide and the activation of peroxidase enzymes leading to the reduction of ascorbic acid and its regenerative action towards lycopene. Tomato purée subjected to sonication at ambient temperature did not change significantly the levels of lycopene (within a 95% of confidence level). The increase in ultrasound frequency (from 18 to 25 kHz) and in power density (from 55 to 5000 W/L) did not affect the concentration of lycopene in tomato purée. However, the levels of lycopene in tomato purée decreased as the process temperature increased. These authors pointed out that ultrasound processing may improve the bioavailability of lycopene by breaking down cell walls, which weakens the bond between lycopene and the fruit tissue. Three main causes could be related to the degradation of lycopene during thermo-sonication: plant stress, oxidation by ascorbic acid and reactions with by-products produced by the peroxidases activity (Oliveira et al., 2015).

Aroma and Sensory Profile

In the study by Šimunek et al. (2013) the effect of high power ultrasound and pasteurization on the aroma profile and sensory properties of apple juice and nectar were investigated. Samples were treated according to the experimental design, with high power sonicator at ultrasound frequency of 20 kHz under various conditions (treatment time: 3, 6 and 9 min, sample temperature: 20, 40 and 60°C, and amplitude: 60, 90 and 120 μm). The aromatic profiles of juices showed that, compared to the untreated samples of juices and nectars, ultrasonic treatment led to the formation of new compounds (which were not present in the untreated samples) or to the disappearance of compounds that were found in the untreated samples. Samples treated at the highest amplitude (120 μm) were used for evaluation and comparison with untreated and pasteurized samples using electronic tongue study. Principal component analysis confirmed the results of electronic tongue study, which showed that the ultrasound-treated and pasteurized juices had different scores compared to the untreated samples.

Apple juice was subjected to ultrasound treatment at 30, 60 and 90% in study by Abid et al. (2014). Amplitude levels using probe-type sonicator (20 kHz, 20°C for 3 min) keeping pulse durations of 5 s on/off, and the samples were stored at 4C for 30 days. Sensory characteristics were investigated. The consumer panel rated sonicated apple juice samples as the most acceptable. Besides, the sensory characteristics were significantly affected during storage period.

With regards to ultrasound-assisted extraction of compounds, it has been shown that ultrasound could be used to extract aromatic chemicals, which imparts bouquet to the wines, over a period of many years (Cocito et al., 1995).

A solvent mixtures of n-pentane and diethyl-ether (1:2) and dichloromethane were used to study the optimization of sonication extraction process. This study emphasized that UAE improved extraction efficiency with increased reproducibility of most aroma compounds than the conventional extraction (Hernanz et al., 1999).

Wang and Zuo (2011) stated in their study of an ultrasonic-assisted hydrolysis of phenolic compounds in cranberry fruits, juice, and sauces samples that ultrasound can speed up the hydrolysis of conjugated phenolics in cranberry products and shorten the hydroxylation time from 16 h required by conventional hydrolysis technique to less than 1.5 h. The analysis time is considerably reduced and twenty phenolic compounds, including benzoic, o-hydroxybenzoic, p-hydroxybenzoic, trans-cinnamic, m-hydroxybenzoic, p-hydroxybenzoic, p-hydroxyphenylacetic, o-phthalic, 2,3-dihydroxybenzoic, vanillic, hydroxycinnamic, 2,4-dihydroxybenzoic, p-coumaric, ferulic, caffeic, and sinapic acids, trans-resveratrol, (-) -epicatechin, catechin, quercetin, and myricetin, have been simultaneously identified and quantified on the basis of GC retention times and simultaneously recorded mass spectra.

Processing of Residues (Waste) from Fruit and Vegetable Juice Processing

The fruit and vegetable industry typically generates large volumes of effluents and solid wastes. The effluents contain high organic loads, cleansing and blanching agents, salt, and suspended solids such as fibers and soil particles. The main solid wastes are organic materials, including discarded fruits and vegetables.

Grape marc is the solid waste of the wine-making process. Consisting of skins, seeds, and small amount of leaves, grape marc has long been used for alcohol, tartaric acid and more recently, the recovery of phenolic compound. Phenolic compounds are of particular interest in wine industry as it gives the characteristics color and flavor to wine, and in pharmaceutical industry for its benefits on human health (Brenna et al., 1998). Results of ultrasonically treated Shiraz and Sangiovese grape marc showed 17 and 35% increase in phenolic compounds respectively, However extraction of these compounds yielded much higher recovery from their respective seeds (Vilkhu et al., 2008).

Tartaric acid is a by-product in the wine industry since a tremendous amount of tartaric acid from lees has to be removed from the wine after yeast fermentation. Tartaric acid is widely used in bakery operations, wine production, pharmaceutical industry, hardening of gypsum, confectionery processing and chemical industry. Palma and Barroso (2002) optimized the ultrasound-assisted extraction conditions for the recovery of tartaric and malic acid from red and white variety grapes purely for quantitative determination in winemaking by-products, as well as Vilkhus group reported an increase 16 to 23% from two different varieties from red grape marc yielded (Vilkhu et al., 2008).

Nephelium lappaceum L. (rambutan) belongs to the family *Sapindaceae*, which is common in Southeast Asia. Due to its refreshing flavor and exotic appearance, this fruit is consumed as fresh, canned or processed, but after being processed, the residues such as seeds and peels were discarded. From the previous findings, it was found that, the discarded peel of rambutan fruit comprises powerful phenolic antioxidants and scavenging activities (Khonkar et al., 2010; Okonogi et al., 2007; Palanisamy et al., 2008; Thitilertdecha et al., 2010) and possess various biological properties such as antibacterial and antiherpes (Nawawi et al., 1999; Thitilertdecha et al., 2010). In the study of

Prakash Maran et al. (2013) the optimum conditions of ultrasound extraction were reported, which were the temperature of 50°C, ultrasound power of 20 W, extraction time of 20 min and solid–liquid ratio of 1.0:18.6. Under these conditions, the predicted maximum yields of total anthocyanins, total phenolics and total flavonoids were 10.71 mg/100 g, 546.98 mg/100 g and 100.93 mg/100 g, respectively.

Large amounts of coconut (*Cocos nucifera*) water and coconut products are processed and consumed in Brazil and many countries. Coconut shell is composed mainly of lignin and cellulose being suitable as a phenolic source. Phenolics from coconut shell are similar to the woods used to store and age alcoholic beverages. Nowadays, the use of coconut shell waste was only applied in agriculture and in some industries which uses coconut fiber. Coconut shell powder is a suitable phenolic source. Rodrigues et al. (2008) reported how the optimum operating condition that maximizes the extraction of phenolic compounds by ultrasound was found using a 50% (v/v) ethanol: water solution, pH adjusted to 6.5, and solvent to solid ratio of 50, being the process carried out at 30°C. The process required 15 min of sonication at ultrasound intensity of 4870 W/m^2. Under this condition 22.44 mg/g of phenolics were extracted from the coconut shell.

The domestic and industrial use of large quantities of citrus fruits, especially for the production of juice, results in the accumulation of high amounts of by-products such as peel, seed, cell and membrane residues which account for about half of the fruit weight. These by-products can be used for the production of molasses, pectins, essential oils, limonene and cattle feed (Bocco et al., 2006). In addition, citrus by-products are a good source of phenolic compounds, especially the characteristic flavanone glycosides which mainly include naringin, hesperidin, narirutin, and neohesperidin (Khan et al., 2010). The ultrasound-assisted extraction of phenolic antioxidants from orange peels with ethanol–water mixtures was proved to be very effective in comparison to conventional procedure. Khan et al. (2010) pointed out the sonication power as the most influential factor in the ultrasound-assisted extraction process followed by temperature and ethanol:water ratio. Ultrasound-assisted extraction can be regarded as an ‘environment-friendly’ or ‘green’ technique for the preparation of extracts rich in natural antioxidants aimed at replacing synthetic antioxidants.

Ultrasound Pre-Treatment Method (Drying, Freezing) in Juice Production

In the study by Villamiel et al. (2015), the influence of power ultrasound on the quality of fruits and vegetables during both the ultrasound pre-treatment and drying has been evaluated. Application of ultrasound in drying of carrots and strawberries led to significant reductions in processing time, while providing high quality end-products. It is known that the most porous materials are more likely to suffer the "sponge effect" caused by the ultrasound and the obtainment of high quality products is mainly due to the acceleration of the drying process. The most recent studies are focused to avoid the heating during dehydration process to minimize the losses of thermolabile compounds. That is the case of the application of ultrasound during freeze-drying (Santacatalina et al., 2015).

Freezing is a widely used process in the food industry for extending the shelf-life of the products due to decreasing the food temperature. The aim of study by Janiszewska and Sakowski (2013) was to investigate a 30-minute ultrasound treatment on the freezing process of carrot juices (9, 12 and 21°Brix). Freezing was conducted by immersion and air chilling method at -30ºC medium temperature. The study examined how ultrasound affects the extract, density of juices, the specific freezing time, freezing point, etc. It was observed that 30-minute ultrasonic treatment did not affect physical properties of tested juices. Solely in the case of higher concentrated juices, the increase of tested parameters was observed. Regardless the concentration or the ultrasound pre-treatment, it has been observed that the specific time required to freeze the product in the immersion method was shorter than in the shock freezing. As a result, the freezing point of carrot juices, after the application of the ultrasound, slightly decreased.

References

Aadil, R. M., Zeng, X.-A., Han, Z. and Sun, D.W. (2013). Effects of ultrasound treatments on quality of grapefruit juice. *Food Chemistry*, *141*(3), 3201–6. doi:10.1016/j.foodchem.2013.06.008.

Aadil, J., Rupasinghe, H. P. V. (2012) Nutritional, Physicochemical and Microbial Quality of Ultrasound-Treated Apple-Carrot Juice Blends. *Food and Nutrition Sciences*, 3, 212–218 doi:10.4236/fns.2012.32031.

Abid, M., Jabbar, S., Wu, T., Hashim, M.M., Bing Hu, Muhammad Saeeduddin and Xiaoxiong Zeng (2014) Qualitative assessment of sonicated apple juice during storage. *Journal of food processing and preservation*, *39*(3), 1745-4549.

Adekunte, A. O., Tiwari, B. K., Cullen, P. J., Scannell, a. G. M. and O'Donnell, C. P. (2010). Effect of sonication on color, ascorbic acid and yeast inactivation in tomato juice. *Food Chemistry*, *122*(3), 500–507. doi:10.1016/j.foodchem.2010.01.026.

Bhat, R., Kamaruddin, N. S. B. C., Min-Tze, L. and Karim, a. a. (2011). Sonication improves kasturi lime (Citrus microcarpa) juice quality. *Ultrasonics Sonochemistry*, *18*(6), 1295–1300. doi:10.1016/j.ultsonch.2011.04.002.

Bocco, A., Cuvelier, M.-E., Richard, H. and Berset, C. (1998). Antioxidant Activity and Phenolic Composition of Citrus Peel and Seed Extracts. *Journal of Agricultural and Food Chemistry*, *46*, 2123–2129. doi:10.1021/jf9709562.

Brenna, O., Buratti, S., Cosio, M. S. and Mannino, S. (1998). A New HPLC Method for the Determination of Polyphenols in Wines Based on the Use of Less Aggressive Eluents and a Coupled Revelation System. *Electroanalysis*, *10*(17), 1204–1207. doi:10.1002/(sici)1521-4109(199811)10:17<1204::aid-elan1204>3.0.co;2-i.

Cheng, L. H., Soh, C. Y., Liew, S. C. and Teh, F. F. (2007). Effects of sonication and carbonation on guava juice quality. *Food Chemistry*, *104*(4), 1396–1401. doi:10.1016/j.foodchem.2007.02.001.

Cocito, C., Gaetano, G. and Delfini, C. (1995). Rapid extraction of aroma compounds in must and wine by means of ultrasound. *Food Chemistry*, *52*(3), 311–320. doi:10.1016/0308-8146(95)92830-D.

Dubrović I., Herceg Z., Režek Jambrak A., Badanjak M. and Dragović-Uzelac V. (2011). Effect of high intensity ultrasound and pasteurization on anthocyanin content in strawberry juice. *Food Technology and Biotechnology,.49*(2), 196–204.

Gogate, P R, and Kabadi, A M (2009) A review of applications of cavitation in biochemical engineering/biotechnology. Biochemical Engineering Journal, *44*(1), 60–72.

Khandpur, Paramjeet; Gogate, Parag R; National Library of Medicine (2015) Effect of novel ultrasound based processing on the nutrition quality of different fruit and vegetable juices. *Ultrasonics sonochemistry, 27*, 125–136.

Hart, E. J. and Henglein, A. (1985). Free radical and free atom reactions in the sonolysis of aqueous iodide and formate solutions. *The Journal of Physical Chemistry*, *89*(20), 4342–4347. doi:10.1021/j100266a038.

Hernanz Vila, D., Heredia Mira, F. J., Beltran Lucena, R. and Fernández Recamales, M. a. (1999). Optimization of an extraction method of aroma compounds in white wine using ultrasound. *Talanta*, *50*(2), 413–421. doi:10.1016/S0039-9140(99)00128-9.

Jeong, J., Kim, J. Y. and Yoon, J. (2006). The role of reactive oxygen species in the electrochemical inactivation of microorganisms. *Environmental Science and Technology*, *40*, 6117–6122. doi:10.1021/es0604313.

Khandpur, P., Gogate, P.R. (2015). Effect of novel ultrasound based processing on the nutrition quality of different fruit and vegetable juices. *Ultrasonic Sonochemistry*. 27, 125-136. doi: 10.1016/j.ultsonch. 2015.05.008.

Khan, M. K., Abert-Vian, M., Fabiano-Tixier, A. S., Dangles, O. and Chemat, F. (2010). Ultrasound-assisted extraction of polyphenols (flavanone glycosides) from orange (Citrus sinensis L.) peel. *Food Chemistry*, *119*(2), 851–858. doi:10.1016/j.foodchem.2009.08.046.

Khonkarn, R., Okonogi, S., Ampasavate, C. and Anuchapreeda, S. (2010). Investigation of fruit peel extracts as sources for compounds with antioxidant and antiproliferative activities against human cell lines. *Food and Chemical Toxicology*, *48*(8-9), 2122–2129. doi:10.1016/j.fct. 2010.05.014.

Janiszewska, E., Sakowski, P. (2013) Effect of the ultrasound on the carrot juices freezing process. *Agricultural engineering*. *4*(148), 49–57.

Long, Jing-jing, Fu, Yu-jie, Zu, Yuan-gang, Li, Ji, Wang, Wei, Gu, Cheng-bo, Meng Luo (2011) Ultrasound-assisted extraction of flaxseed oil using immobilized enzymes *Bioresource Technology,* 102, 9991–9996.

Lou, Z., Wang, H., Li, J., Zhu, S., Lu, W. and Ma, C. (2011). Effect of simultaneous ultrasonic/microwave assisted extraction on the antioxidant and antibacterial activities of burdock leaves, *5*(22), 5370–5377.

Martínez-Flores, Héctor E., Guadalupe Garnica-Romo, Ma., Bermúdez-Aguirre, Daniela.

Raj Pokhrel, Prashant, Barbosa-Cánovas, Gustavo V. (2015) Physico-chemical parameters, bioactive compounds and microbial quality of thermo-sonicated carrot juice during storage. *Food Chemistry, 172*, 650–656.

Mason, T. (1996). The uses of ultrasound in food technology. *Ultrasonics Sonochemistry*, *3*(3), S253–S260. doi:10.1016/S1350-4177(96)00034-X.

Nawawi, A, Nakamura, N., Hattori, M., Kurokawa, M. and Shiraki, K. (1999). Inhibitory effects of Indonesian medicinal plants on the infection of herpes simplex virus type 1. *Phytotherapy Research : PTR*, *13*(1), 37–41. doi: 10.1002/(SICI)1099-1573(199902)13:1<37:: AID-PTR382>3.0.CO;2-S.

Okonogi, S., Duangrat, C., Anuchpreeda, S., Tachakittirungrod, S. and Chowwanapoonpohn, S. (2007). Comparison of antioxidant capacities and cytotoxicities of certain fruit peels. *Food Chemistry*, *103*(3), 839–846. doi:10.1016/j.foodchem.2006.09.034.

Oliveira, Valéria S., Rodrigues, Sueli, Fernandes, Fabiano A.N. (2015) Effect of high power low frequency ultrasound processing on the stability of lycopene. *Ultrasonics Sonochemistry, 27*, 586–591.

Palanisamy, U., Cheng, H. M., Masilamani, T., Subramaniam, T., Ling, L. T. and Radhakrishnan, A. K. (2008). Rind of the rambutan, Nephelium lappaceum, a potential source of natural antioxidants. *Food Chemistry*, *109*, 54–63. doi:10.1016/j.foodchem.2007.12.018.

Palma, M. and Barroso, C. G. (2002). Ultrasound-assisted extraction and determination of tartaric and malic acids from grapes and winemaking by-products. *Analytica Chimica Acta*, *458*, 119–130. doi:10.1016/S0003-2670(01)01527-6.

Prakash Maran, J., Manikandan, S., Vigna Nivetha, C. and Dinesh, R. (2013). Ultrasound assisted extraction of bioactive compounds from Nephelium lappaceum L. fruit peel using central composite face centered response surface design. *Arabian Journal of Chemistry*. in press. doi:10.1016/ j.arabjc.2013.02.007.

Režek Jambrak, A. and Herceg, Z. (2014) Application of Ultrasonics in Food preservation and processing. In:Conventional and Advanced Food Processing Technologies, Suvendu Bhattacharya (ed.). UK, John Wiley and Sons, Ltd., pp.515-535.

Rodrigues, S., Pinto, G. a S. and Fernandes, F. a N. (2008). Optimization of ultrasound extraction of phenolic compounds from coconut (Cocos nucifera) shell powder by response surface methodology. *Ultrasonics Sonochemistry*, *15*(1), 95–100. doi:10.1016/j.ultsonch. 2007.01.006.

Rostagno, M. a., Palma, M. and Barroso, C. G. (2003). Ultrasound-assisted extraction of soy isoflavones. *Journal of Chromatography A*, *1012*(2), 119–128. doi:10.1016/S0021-9673(03)01184-1.

Rostagno, M. A., Palma, M. and Barroso, C. G. (2007). Ultrasound-assisted extraction of isoflavones from soy beverages blended with fruit juices. *Analytica Chimica Acta*, *597*(2), 265–272. doi:10.1016/j.aca.2007.07.006.

Safari, Masoumeh, Ghanati, Faezeh, Mehrdad Behmanesh, Abazar Hajnorouzi, Bahareh Nahidian, Ghahremani Mina (2013) Enhancement of antioxidant enzymes activity and expression of CAT and PAL genes in hazel (Corylus avellana L.) cells in response to low-intensity ultrasound. *Acta Physiologiae Plantarum 35*(9), 2847–2855. DOI 10.1007/s11738-013-1318-6.

Shiferawterefe, N, Buckow,R., Versteeg, C. (2015) Quality-Related Enzymes in Plant-Based Products: Effects of Novel Food-Processing Technologies Part 3. Ultrasonic Processing. *Critical Reviews in Food Science and Nutrition*, *55*, 147–158.

Subhedar, Preeti B., Babu, Narmadha R., Gogate, Parag R. (2015) Intensification of enzymatic hydrolysis of waste newspaper using ultrasound for fermentable sugar production *Ultrasonics Sonochemistry*. 22, 326–332.

Šimunek, M., Jambrak Režek, A., Petrović, M., Juretić, H., Major, N., Herceg, Z., Hruškar, M. and Tomislava Vukušić (2013). Aroma Profile and Sensory Properties of Ultrasound-Treated Apple Juice and Nectar. *Food Technology and Biotechnology*, *51*(1), 101–111.

Thitilertdecha, N., Teerawutgulrag, A., Kilburn, J. D. and Rakariyatham, N. (2010). Identification of Major Phenolic Compounds from Nephelium lappaceum L. and Their Antioxidant Activities. *Molecules*, *15*, 1453–1465. doi:10.3390/molecules15031453.

Tiwari, B. K., O'Donnell, C. P., Patras, A. and Cullen, P. J. (2008). Anthocyanin and ascorbic acid degradation in sonicated strawberry juice. *Journal of Agricultural and Food Chemistry*, *56*(21), 10071–10077. doi:10.1021/jf801824v.

Tiwari, B. K., O'Donnell, C. P. and Cullen, P. J. (2009). Effect of non thermal processing technologies on the anthocyanin content of fruit juices. *Trends in Food Science and Technology*, *20*(3–4), 137–145. doi:10.1016/j.tifs.2009.01.058.

Valero, M., Recrosio, N., Saura, D., Muñoz, N., Martí, N. and Lizama, V. (2007). Effects of ultrasonic treatments in orange juice processing. *Journal of Food Engineering*, *80*(2), 509–516. doi:10.1016/j.jfoodeng.2006.06.009.

Vilkhu, K., Mawson, R., Simons, L. and Bates, D. (2008). Applications and opportunities for ultrasound assisted extraction in the food industry - A review. *Innovative Food Science and Emerging Technologies*, *9*(2), 161–169. doi:10.1016/j.ifset.2007.04.014.

Villamiel, M., Juliana Gamboa, A. Cristina Soria, Enrique Riera, José V. García-Pérez, Antonia Montilla (2015) Impact of power ultrasound on the quality of fruits and vegetables during dehydration. *Physics Procedia,* 70, 828–832.

Wang, C. and Zuo, Y. (2011). Ultrasound-assisted hydrolysis and gas chromatography-mass spectrometric determination of phenolic compounds in cranberry products. *Food Chemistry*, *128*(2), 562–568. doi:10.1016/j.foodchem.2011.03.066.

Zenker, M., Heinz, V. and Knorr, D. (2003). Application of Ultrasound-Assisted Thermal Processing for Preservation and Quality Retention of Liquid Foods. *Journal of Food Protection*, *66*(9), 1642–1649.

In: Applications of Ultrasound in the Beverage… ISBN: 978-1-63485-069-8
Editor: Juan Francisco García Martín

Chapter 2

STATE OF THE ART OF THE USE OF ULTRASOUND IN THE BEVERAGE INDUSTRY (II): EFFECTS ON MICROORGANISMS

Anet Režek Jambrak*[*], *Tomislava Vukušić* and *Zoran Herceg

Faculty of Food technology and Biotechnology,
Zagreb, Croatia

ABSTRACT

Ultrasonication is a nonthermal method of food processing that has the advantage of preserving fruit juices without/or minor causing the common side effects associated with conventional heat treatments. Sonication technology can improve the process through reduced processing time, higher throughput and lower energy consumption. When ultrasound is used in inactivation of microorganisms, it would have to be used in conjunction with pressure treatment (manosonication), heat treatment (thermosonication) or both (manothermosonication). The effect of ultrasound can be physical (cavitation, mechanical effects, micromechanical shocks) or chemical due to formation of free radicals (H· and OH· due to sonochemical reaction) formed by the decomposition of water inside the oscillating bubbles. Many researchers have demonstrated excellent effects in inactivation of microorganisms in food.

[*] Corresponding Author Email: arezek@pbf.hr.

Today, consumers make demands on the quality, flavor and taste of different kinds of fruit juices. The process of inactivation of microorganisms in fruit juices should be effective in inactivation of microorganisms and retaining quality, sensory, nutritional and organoleptic factors of fruit juices.

Keywords: fruit juices, inactivation, microorganisms, beverages

INTRODUCTION

Ultrasound is the sound that is above the threshold of the human ear (above 18 kHz).

Ultrasound is typically divided into three regions of frequency. Power ultrasound is in the region from 16–100 kHz (1 Hertz is 1 cycle/sec), high-frequency ultrasound is from 100 kHz–1 MHz and diagnostic ultrasound is from 1–10 MHz. Ultrasound is generated with either piezoelectric or magnetostrictive transducers that create high-energy vibrations.

Power ultrasound (20-100 kHz) can provide the mechanical effect of cavitation in liquid systems which can alter physical and chemical properties of food depending on the type of material involved. When ultrasound waves pass through a medium, a series of compression and rarefaction waves on the molecules of the medium are produced. This enforces a sinusoidal acoustic pressure (Pa) in addition to the hydrostatic pressure acting on the medium (Režek Jambrak and Herceg, 2014). If a large negative pressure (sufficiently below ambient) is applied to the liquid so that the distance between the molecules exceeds the critical molecular distance necessary to hold the liquid intact, the liquid will break down and the cavitation bubbles will be formed. These bubbles are formed from the gas nuclei within the fluid and are distributed throughout the liquid. After a period of a few cycles, the bubbles will grow into a critical size which makes them unstable and collapse violently (Režek Jambrak and Herceg, 2014). The extent of cavitation is affected by energy, intensity, medium viscosity, surface tension, vapor pressure, nature and concentration of dissolved gas, temperature and pressure of the treatment and the presence of solid particles. Two competing theories exist to explain the chemical effects due to cavitation - the *hot-spot theory* and the *electrical theory*. The hot-spot theory postulates that when the bubbles cavitate, localized hot spots are formed which reach temperatures and pressures in excess of 5000 K and 1000 atm, respectively. Under these extreme conditions, chemical compounds can be degraded through three distinct pathways such as (1) oxidation by hydroxyl radicals, (2) pyrolytic decomposition, and (3) supercritical water oxidation. The electrical theory postulates that an electrical charge is created on the surface of cavitation bubble, forming enormous electrical field gradients across the bubbles which are capable of bond breakage upon collapse.

Pasteurization of fruit juices is very effective against pathogens and several spoilage microorganisms. Thermal processing affects sensory and nutritional properties. Demands for nutritious and safe foods have resulted in increased interest in nonthermal preservation techniques. The *U.S. Food and Drug Administration* (FDA) requests the potential to meet 5 log microbial reductions, in order to have safe product. Power ultrasound is technique that is used in processing of fruit and vegetable juices. Several mechanisms of inactivation of microorganisms have been proposed. The mechanical stress of ultrasound on cells suspended in liquid medium (hydrodynamic events: acoustic cavitation and cavitation induced microstreaming) and radical formation are proposed mechanisms for microbial inactivation. Physical and chemical factors affect fruits, vegetables and plant tissues by causing the excessive accumulation of reactive oxygen species (ROS). Reactive oxygen species can cause damage to membranes and other essential macromolecules such as proteins, DNA, and lipids.

CASE STUDIES

Fruit juices are spoiled primarily because of growth of acid tolerant and osmophilic microorganisms. Thermal pasteurization is commonly used in fruit juice industries for the preservation of fruit juices. Due to thermal treatment, there are changes in essential nutrients and in physicochemical and organoleptic properties. Nonthermal pasteurization methods such as power ultrasound have been employed in fruit juices processing to overcome the negative effects of thermal pasteurization. Power ultrasound has been identified as potential technology to meet the US Food and Drug Administration (FDA) requirement of 5 log reduction of *Escherichia coli* in fruit juices by thermosonication treatment (Lee et al., 2013). High power ultrasound causes bubble cavitation in a liquid due to pressure changes. These resultant microbubbles collapse violently in the succeeding compression cycles of propagated ultrasonic waves resulting in localized high temperature up to 5000 K, pressures up to 50,000 kPa, and high shearing effects causing breakdown of cell walls, disruption of cell membranes, and damage of DNA of microorganisms (Aneja et al., 2014). The simultaneous application of ultrasound and mild heat in fruit juice processing industry has the greatest potential and benefits of juice quality preservation. For a shorter processing time, it can be categorized as minimal processing for freshness and health purposes (Abdullah and Chin, 2014).

Generally, cavitation is more effective on gram-positive bacteria, spores, spherical-shaped and small round cells (Režek Jambrak and Herceg, 2014). Gram-positive bacteria are more resistant to ultrasound than gram-negative because of their thicker cells wall (tightly adherent layer of peptidoglycans) which gives them a better protection against ultrasound effects. Bacterial spores and fungi are more resistant to ultrasound than vegetative bacteria. Spores are more difficult to be destroyed than vegetative cells which are in phase of growth.

In study of Abid et al. (2013) the effects of combined treatment of ultrasound (US) and high hydrostatic pressure (HHP) on microorganisms (total plate counts, yeasts and molds) were examined. The reduction of microbial populations in US treated samples was due to the cavitation that caused damage to biological species by increasing localized temperature and pressure of the product. These created shock waves and hydroxyl radicals, which ultimately enhanced biocidal effect (Abid et al., 2013). The higher inactivation of microbial cells in the combined US-HHP treatments in the cited study might be due to the pronounced and additive effects of both techniques.

Bevilacqua et al. (2013) reported the use of ultrasound (US) as a suitable strategy to control the growth of spoiling yeasts in fruit juices. In a first phase, US technique was tested towards *Saccharomyces cerevisiae* inoculated in different juices (strawberry, orange, apple, pineapple and red-fruits): the treatment was performed by modulating the level of the power (20–60%), the duration of the treatment (2–6 min) and the pulse (2–6 s), according to a fractional design. Then, the best treatment was applied against some other spoiling yeasts (*Pichia membranifaciens*, *Wickerhamomyces anomalus*, *Zygosaccharomyces bailii*, *Zygosaccharomyces rouxii*, *Candida norvegica*). The results showed that the effect of US was mainly influenced by the power and the duration of the treatment. The highest reduction of *S. cerevisiae* was found in the following combinations of the design: power 60%, time 4 min and pulse 2 s, and power 60%, time 6 min and pulse 6 s; these results were confirmed for the other spoiling yeasts. US and citrus extract used could be combined to prolong the shelf life of the red-fruit juice and control the growth of *Z. bailii*.

In study by Jambrak et al. (2013) (unpublished data) the effect of non-thermal technology, high power ultrasound (HPU), on inactivation of *Aspergillus ochraceus* 318, *Penicillium expansum* 565, *Rhodotorula* sp. 74, *Saccharomyces cerevisiae* 5 and *Alicyclobacillus acidoterrestris* DSM 3922 in clear juices and nectars from apple, blueberry and cranberry juice concentrate was investigated. Inoculated juice and nectars were treated by high power ultrasound (20 kHz) according to procedure set by central composite design (CCD). Three operational parameters, amplitude (60, 90 and 120 μm), temperature (20, 40 and 60°C), and treatment time (3, 6 or 9 min) were varied in order to observe the influence of ultrasound and combination of ultrasound and slight heating (thermosonication) on growth and inactivation of selected microorganisms. Number of vegetative cells of *A. acidoterrestris* DSM 3922 were not significantly reduced by high power ultrasound ($p > 0.05$), except in apple juice, where statistical significant ($p < 0.05$) influence of quadratic interaction of amplitude on bacteria reduction was observed. In all samples of fruit juices and nectars in terms of ultrasonic treatment at 60°C and times of 3, 6 and 9 minutes, regardless of the value of the amplitude, complete inactivation of the growth of yeasts and moulds were achieved, while at 20 and 40°C it was not observed. The value of reduction of cells of selected yeasts and moulds for ultrasound treatments at 60°C and the duration of the 3, 6 and 9 minutes ranged from 3.556 to 5.934 log units, depending on the initial number of selected yeasts and moulds before treatment.

The effect of ultrasonic treatments on *Alicyclobacillus acidoterrestris* in apple juice was investigated by Yuan et al. (2009). In general, inactivation of the cells was more pronounced at an elevated power level and as the processing time increased. Approximately 60% of the cells were inactivated after treating the apple juice with 300W ultrasound for 30 min. The reduction reached more than 80% when the juice was processed for 60 min. The linear inactivation rates (D values) of ultrasound on *A. acidoterrestris* were both process- and strain dependent. The lowest D value at 36.18 min was found when using 600W ultrasound to treat the A30 strain isolated from the air of a commercial apple juice processing plant, whereas the strain isolated from apple juice concentrate was found most resistant against ultrasound. Significant reduction of different strains of thermoacidophilic *A. acidoterrestris* in apple juice was achieved by subjecting the apple juice to ultrasonic treatments operated at 23 kHz between power levels of 400 and 600 W and processing time from 30 to 60 min. Although ultrasound might not be able to serve as a stand-alone unit operation to inactivate *A. acidoterrestris*, it could be employed as a secondary operation following thermal inactivation to ensure effective destruction of residual cells.

Adekunte et al. (2010) studied the effect of sonication on the resistance of yeast (*Pichia fermentans*) in tomato juice. Tomato juice samples were sonicated at amplitude levels ranging from 24.4 to 61.0 μm at a constant frequency of 20 kHz for different treatment times and pulse durations. Significant reductions were observed at higher amplitudes and processing times. Yeast inactivation was found to follow the Weibull model with a high regression coefficient ($R^2 > 0.98$). The desired 5 log reductions (D5 value) and shape factors were found to correlate exponentially with amplitude level. Results presented in this study show that sonication alone is an effective process to achieve the desired level of yeast inactivation in tomato juice.

Char et al. (2010) studied the potential application of ultrasound to juices and evaluated the survival of *Escherichia coli*, *Saccharomyces cerevisiae*, and a yeast cocktail in orange and apple juices and in 0.1% w/w peptone water. They used high-intensity ultrasound (USc) and/or short-wave ultraviolet radiation (UV-C). Combined treatment was more effective in simultaneous rather than in a series of USc−UV-C arrangement. The application of high-intensity ultrasound in a continuous flow mode had low efficacy for all studied microorganisms inoculated in peptone water, orange, or apple juices.

Carrot juice was thermo-sonicated (24 kHz, 120 μm amplitude) at 50°C, 54°C and 5°C for 10 min. Samples sonicated at 58°C had the best quality; microbial growth remained low at around 3-log for mesophiles, 4.5-log for yeasts and molds and 2-log for enterobacteria after 20 d of storage. Furthermore, thermo-sonicated juice at 58°C retained >98% of carotenoids and 100% of ascorbic acid. The influence of sonication is very clear, since during the cavitation some free radicals are also produced, resulting in an antimicrobial effect through the storage (Martínez-Flores et al., 2015).

Wine lees that had spent more than one year in oak barrels were used for ultrasound-assisted yeast lysis (22 W/L, 18ºC) in a model wine and the obtained results were compared with those of a classical yeast autolysis at 25ºC (García Martín et al., 2013). Both classical autolysis and ultrasound-assisted lysis increased the concentration of carbohydrates in the model media, but only ultrasound markedly increased the release of proteins. As for yeast viability, classical autolysis did not have noticeable effect on yeasts. On the contrary, ultrasound provoked a high cell disruption, and after 20 h ultrasonic treatment, yeast hardly grew in any agar assayed. The inactivation rate constant of ultrasound-assisted yeast lysis was 2.54×10^{-5} s^{-1}. Furthermore, the cell concentration at the end of the ultrasound-assisted yeast lysis was much lower than that of the classical autolysis. The morphological changes in cells observed by scanning electron microscopy confirmed this effect of ultrasound on yeast cells. On the other hand, the cell disruption provoked by ultrasound was related to the increase of the protein content, since complete cell inactivation was found at the end of the exponential increase phase of protein release to the model wine. Finally, these authors found that the effect of ultrasound on *Saccharomyces* and *Brettanomyces* spp. cells, the most common yeasts involved in winemaking, was the same (García Martín et al., 2013).

Gabriel (2014) established the inactivation kinetic parameters of some pathogenic bacteria including *Escherichia coli* O157:H7, *Salmonella enterica serotypes*, and *Listeria monocytogenes*; and spoilage yeasts namely, *Debaryomyces hansenii*, *Clavispora lusitaniae*, *Torulaspora delbrueckii*, *Pichia fermentans*, and *Saccharomyces cerevisiae* in orange juice. Juice with microorganisms was subjected to multi-frequency Dynashock power ultrasound treatment. All test organisms exhibited a biphasic inactivation behavior with a sigmoidal inactivation curve consisted of an initial inactivation lag, followed by logarithmic linear inactivation. Injury accumulation in the inactivation lag phase was established in acid-adapted bacteria. The time necessary to reduce initial inoculated populations by 5 log cycles (99.999%), T5D values, significantly increased with acid adaptation. The T5D of *E. coli*, *S. enterica*, and *L. monocytogenes* increased from 37.64, 36.87, and 34.59, respectively; to 54.72, 40.38, and 37.83 min, respectively, after acid exposure. Temperature increase due to heat propagation during ultrasound treatment decreased the resistance of the test bacteria. The cocktail of *E. coli* O157:H7 had significantly greater resistance towards ultrasound treatment (T5D = 54.72 min) than any of the individual strain (T5D = 41.48–47.48 min) in the mix. Similar results were found in the composited (T5D = 60.02 min) and individual species (T5D = 20.31–59.04 min).

Microbial and enzymatic inactivation in fresh pear juice was studied by Saeeduddin et al. (2015) under conventional and ultrasound-combined pasteurizations. The ultrasound-pasteurization conditions were 25, 45 and 65ºC for 10 min using a 750 W probe sonicator (frequency 20 kHz and amplitude 70%) while conventional pasteurization conditions were 65ºC for 10 min and 95ºC for 2 min. Complete inactivation of microbes (total plate count, yeast and mould) were observed in conventional pasteurization at 95°C but this treatment also showed highest losses of ascorbic acid, total phenols, flavonoids and antioxidant capacity. Ultrasound-pasteurization at 65ºC for 10 min showed the best results in retention of ascorbic acid and other phenolic compounds along with complete inactivation of microbes.

Ortuño et al. (2012) studied the effect of ultrasound on the SC-CO_2 inactivation kinetics of *E.coli* at different pressures (100, 225, 290 and 350 bar; 36°C), temperatures (31°C, 36°C and 41°C; 225 bar) and varying the composition of the medium (LB Broth, apple and orange juice). Using only SC-CO_2 the inactivation rate in LB Broth increased progressively as the pressure or temperature rose and the average time needed to achieve a reduction of 8 log-cycles was 50 min. When both SC-CO_2 and ultrasound were used in LB Broth, 95% less time was needed, on average, to achieve the same reduction of 8 log-cycles and the effect of pressure and temperature was minimized; this reduction being achieved after only 2 min at 31°C and 225 bars. Using the ultrasonic system led to a shorter process time regardless of the treatment medium, preventing the inhibitory effect that the solutes from the juices had on the SC-CO_2 inactivation. The technology developed permits both a drastic decrease in SC-CO_2 inactivation times and also the use of mild process conditions.

The objective of the study by Ortuño et al. (2013) was to analyze the influence of high power ultrasound (HPU) on the supercritical carbon dioxide (SC-CO_2) inactivation kinetics of *Saccharomyces cerevisiae* and to determine the effect of the temperature (31–41°C), pressure (100–350 bar) and composition of the medium (YPD Broth, apple and orange juice) on the process of inactivation. Using a batch-mode SC-CO_2 at 350 bar and 36°C, a reduction of 6.7 log-cycles was obtained after 140 min of treatment. However, when HPU (40 W ± 5 W and 30 kHz) was applied during the SC-CO_2 treatments, a reduction of 7 log-cycles was achieved after 2 min of treatment for all pressures and temperatures applied. The effect of increasing pressure (from 100 to 350 bar, 36°C) or temperature (from 31 to 41°C, 225 bar) did not significantly influence this inactivation level. The application of ultrasound leads to a vigorous agitation and cavitation which could accelerate the SC-CO_2 dissolving in the medium. This accelerates the penetration of CO_2 into cells and its inactivation mechanisms. In batch operations the application of HPU increases the speed of reaching saturation solubility of CO_2 in many liquid media and significantly reduces microbial inactivation times.

Ortuño et al. (2014) studied the combined effect of supercritical carbon dioxide (SC-CO_2) and high power ultrasound (HPU) on the inactivation kinetics of *Escherichia coli* and *Saccharomyces cerevisiae* in orange juice. Experiments were performed at different temperatures (31-41°C, 225 bars) and pressures (100-350 bar, 36°C). The inactivation rate of *E. coli* and *S. cerevisiae* increased with pressure and temperature during the HPU-assisted SC-CO_2 treatments. The HPU-assisted SC-CO_2 inactivation kinetics were represented by models that included temperature, pressure and treatment time as variables, based on the Biphasic, the Peleg Type B, and the fractional models. The HPU-assisted SC-CO_2 batch system permits the use of mild process conditions and treatment times that can be even shorter than those of continuous SC-CO_2 systems.

Pineapple, grape and cranberry juice were thermo-sonicated (24 kHz, 400 W, 120 μm) at 40°C, 50°C and 60°C during 10 min at continuous and pulsed mode (Bermúdez-Aguirre and Barbosa-Cánovas, 2012). Inactivation of *Saccharomyces cerevisiae* was tested from 0 to 10 min. Survivor's curves were fitted with Weibull distribution, four parameter model and modified Gompertz equation. The acoustic energy was also calculated. *S. cerevisiae* was inactivated in the treatments at 60°C, with the continuous mode being more effective. Grape juice showed total inactivation (7-log) after 10 min. The modified Gompertz equation showed the best fit. Energy analysis showed that pineapple juice (4287.02 mW/mL) required a higher amount of energy; grape juice showed the lowest value (3112.13 mW/mL). Because of the above-mentioned results, these authors concluded that ultrasound represents a viable option for juice pasteurization (Bermúdez-Aguirre and Barbosa-Cánovas, 2012).

The combined effect of the simultaneous application of heat and ultrasonic waves under pressure (manothermosonication, MTS) on the survival of a strain of *Cronobacter sakazakii* was studied in apple juice (Arroyo et al., 2012). Below 45°C, the inactivation provoked by ultrasound under pressure was independent of temperature. Above 64°C, the lethal effect of ultrasound under pressure was negligible when compared to the lethality of the heat treatment at the same temperature. Between 45°C and 64°C, the lethality of the combined process (MTS) was higher than expected if heat and ultrasound under pressure processes acted simultaneously but independently, that is, a synergistic effect was observed. The maximum synergistic effect (38.2%) was found at 54°C. Recovery on selective media – with sodium chloride or bile salts – revealed that a certain proportion of the survivors after MTS treatments were sublethally injured. It was also observed that survivors after MTS treatments progressively died during refrigerated storage (up to 96 h at 4°C) in the apple juice (Arroyo et al., 2012).

Wong et al. (2010) used osmosonication (ultrasound with nonthermal concentration) on tropical highland blackberry (*Rubus adenotrichus*) juice inoculated with *Salmonella* spp., *Shigella* sp. (a lactic acid bacterium) yeasts, and molds over different periods of time to assess reductions in microorganism. The juice was then sonicated for 5.9 to 34.1 min at 20 kHz and 0.83 W/mL. Nonthermal concentration was simulated by mixing the juice with a concentrate to obtain 650 g total soluble solids (TSS/kg). The lactic acid bacterium, yeasts, and molds were reduced by 1.60 to as much as 5.01 log_{10} CFU/mL, whereas, for pathogens, reductions were total ≥7.1 log_{10} CFU/mL after 24 h of storage.

The aim of study by Ferrario et al. (2015) was to evaluate the effect of ultrasound (600 W, 20 kHz and 95.2 mm wave amplitude; 10 or 30 min at 20, 30 or 44ºC) with pulsed light on the inactivation of *Alicyclobacillus acidoterrestris ATCC 49025* spores and *Saccharomyces cerevisiae KE162* inoculated in commercial and natural squeezed apple juices. Combination of these technologies led up to 3.0 log cycles of spore reduction in commercial apple juice and 2.0 log cycles in natural juice. On the other hand, 6.4 and 5.8 log cycles of reduction were achieved for *S. cerevisiae* in commercial and natural apple juices, respectively.

Samani et al. (2105) studied the effectiveness of combined microwave-ultrasonic pasteurization system on *Escherichia coli* in sour cherry juice. Based on their findings, microwave output power, ultrasound power, and ultrasonic exposure time as well as the microwave-induced temperature were the most effective factors in reducing *E. coli.* In addition, the microwave-induced temperature and ultrasonic exposure time, as independent variables, were both effective on *E. coli* removal. At higher temperatures, the effectiveness of ultrasonic waves as well as cavitation intensity declined. However, the combined effect of ultrasound and temperature was more significant than their individual effect. On the basis of *RSM* modeling, the optimum processing condition was: 352.21W microwave output power, 49.94°C temperature, 475.13W ultrasound power and 6 minutes of exposure time. When the ultrasound power was increased, the amount of *E. coli* was decreased due to the increased intensity of cavitation.

Finally, Pala et al. (2015) studied the effects of ultrasound treatment at various amplitudes (50, 75, and 100%) and times (0, 6, 12, 18, 24, and 30 min) on *Escherichia coli* ATCC 25922 (a surrogate for *E. coli* O157:H7) and *Saccharomyces cerevisiae* ATCC 2366 in pomegranate juice. More than a 5-log inactivation of *E. coli* ATCC 25922 and a 1.36-log inactivation of *S. cerevisiae* ATCC 2366 were achieved after 30 min of ultrasound treatment at 100% amplitude. The log-linear and Weibull models were successfully used to estimate the microbial inactivation as a function of ultrasound treatment time (R^2 = 0.97). US processing achieved more than a 5-log reduction in *E. coli* ATCC 25922 which meets the U.S. Food and Drug Administration guidelines regarding pathogen reduction in fruit juices. The effect of ultrasound treatment on *S. cerevisiae* was more limited, resulting in only a 1.36-log reduction under the maximum processing conditions, which is why additional treatments such as mild heat, pressure, or a natural antimicrobial are required to control yeast in sonicated juice.

CONCLUSION

Methods and results from the above-mentioned studied cases showed that ultrasound treatment is a potential tool for inactivation of microorganisms. Moreover, improved inactivation effect of ultrasound can be observed when combining high power ultrasound (HPU) with elevated temperature (thermosonication), and pressure (manothermosonication). Also, high power ultrasound can be combined with supercritical carbon dioxide to obtain even shorter processing time. Application of ultrasound showed great potential to be used as processing method since this technique leads to operating time reducing, less chemicals usage, low energy consumption and less production of CO_2. Therefore, it is suggested to use term 'green non-thermal food processing technique' for high power ultrasound.

REFERENCES

Aadil, J., Rupasinghe V. (2102) Nutritional, Physicochemical and Microbial Quality of Ultrasound-Treated Apple-Carrot Juice Blends (2012) *Food and Nutrition Sciences*, 3, 212–218 doi:10.4236/fns.2012.32031.

Abid, M., Jabbar, S., Hu, B., Hashim, M.M, Wua, T., Wu, Z., Khan, M. A., Zeng, X. (2014) Synergistic impact of sonication and high hydrostatic pressure on microbial and enzymatic inactivation of apple juice. *LWT - Food Science and Technology*, 59, 70–76.

Abdullah, N., Ling Chin, N. (2014) Application of Thermosonication Treatment in Processing and Production of High Quality and Safe-to-Drink Fruit Juices. “ST26943”, 2nd International Conference on Agricultural and Food Engineering, CAFEi2014, *Agriculture and Agricultural Science Procedia,* 2, 320–327.

Adekunte, A., Tiwari, B.K., Scannell, A., Cullen, P.J., O'Donnell, C. (2010) Modelling of yeast inactivation in sonicated tomato juice. *International Journal of Food Microbiology*, 137, 116–120.

Aneja, K.R, Dhiman, R., Aggarwal, N.K Aneja, A. (2014) Emerging Preservation Techniques for Controlling Spoilage and Pathogenic Microorganisms in Fruit Juices. *Hindawi Publishing Corporation International Journal of Microbiology*, Article ID 758942.

Arroyo, C., Cebrián, G., Pagán, R., Condón, S. (2012) Synergistic combination of heat and ultrasonic waves under pressure for *Cronobacter sakazakii* inactivation in apple juice *Food Control*, 25(1), 342–348. DOI:10.1016/j.foodcont.2011.10.056.

Bermúdez-Aguirre, D., Barbosa-Cánovas, G.V. (2012) Inactivation of Saccharomyces cerevisiae in pineapple, grape and cranberry juices under pulsed and continuous thermo-sonication treatments. *Journal of Food Engineering*, 108(3), 383–392. DOI:10.1016/j.jfoodeng.2011.06.038.

Bevilacqua, A. and Speranza, B. and Campaniello, D. and Sinigaglia, M. and Corbo, M.R (2013) Inactivation of Spoiling Yeasts of Fruit Juices by Pulsed Ultrasound. *Food and Bioprocess Technology*, 7, 2189–2197.

Char, C.D, Mitilinaki, E., Guerrero, S.N, Alzamora, S.M. (2010) Use of High-Intensity Ultrasound and UV-C Light to Inactivate Some Microorganisms in Fruit Juices. *Food and Bioprocess Technology*, 3, 797–803. DOI 10.1007/s11947-009-0307-7.

Ferrario, M., Alzamora, S.M, Guerrero, S. (2015) Study of the inactivation of spoilage microorganisms in apple juice by pulsed light and ultrasound. *Food Microbiology*, 46, 635–642.

Gabriel, Alonzo A. (2014) Inactivation behaviors of foodborne microorganisms in multi-frequency power ultrasound-treated orange juice. *Food Control*. 46, 189–196. DOI:10.1016/j.foodcont.2014.05.012.

García Martín J.F., Guillemet L., Feng C., Sun D.W. (2013) Cell viability and proteins release during ultrasound-assisted yeast lysis of light lees in model wine. *Food Chemistry*, 141, 934–939. DOI:10.1016/j.foodchem.2013.03.081.

Martínez-Flores, H.E., Garnica-Romo, G.A., Bermúdez-Aguirre, D., Pokhrel, P.R., Barbosa-Cánovas, G.V. (2015) Physico-chemical parameters, bioactive compounds and microbial quality of thermo-sonicated carrot juice during storage. *Food Chemistry*, 172, 650–656. DOI:10.1016/j.foodchem.2014.09.072.

Ortuño, C., Martínez-Pastor, M.T., Mulet, A., Benedito, J. (2012) An ultrasound enhanced system for microbial inactivation using supercritical carbon dioxide. *Innovative Food Science and Emerging Technologies*, 15, 31–37. DOI:10.1016/j.ifset.2012.02.006.

Ortuño, C., Martínez-Pastor, M.T., Mulet, A., Benedito, J. (2013) Application of high power ultrasound in the supercritical carbon dioxide inactivation of Saccharomyces cerevisiae. *Food Research International*, 51(2), 474–481. DOI:10.1016/j.foodres.2013.01.041.

Ortuño, C., Balaban, M., Benedito, J. (2014) Modeling of the inactivation kinetics of Escherichia coli, Saccharomyces cerevisiae and pectin methylesterase in orange juice treated with ultrasonic-assisted supercritical carbon dioxide. *Journal of Supercritical Fluids*, 90, 18–26. DOI:10.1016/j.supflu.2014.03.004.

Pala, C.U., Zorba, N.N.D., Zcan, G.I. (2015) Microbial Inactivation and Physicochemical Properties of Ultrasound Processed Pomegranate Juice. *Journal of Food Protection*, 78(3), 531–539.

Režek Jambrak, A. and Herceg, Z. (2014) Application of Ultrasonics in Food preservation and processing. In:Conventional and Advanced Food Processing Technologies, Suvendu Bhattacharya (ed.). UK, John Wiley and Sons, Ltd., pp. 515–535.

Safari, M., Ghanati, F., Behmanesh, M., Hajnorouzi, A., Nahidian, B., Mina, G. (2013) Enhancement of antioxidant enzymes activity and expression of CAT and PAL genes in hazel (*Corylus avellana* L.) cells in response to low-intensity ultrasound. *Acta Physiologiae Plantarum*, 35(9), 2847–2855. DOI:10.1007/s11738-013-1318-6.

Samani, B.H, Khoshtaghaza, M.H, Minaee, S. Abbasi, S. (2015) Modeling the Simultaneous Effects of Microwave and Ultrasound Treatments on Sour Cherry Juice Using Response Surface Methodology. *Journal of Agricultural Science and Technology,* 17, 837–846.

Saeeduddin, M., Abid, M., Jabbar, S., Wu, T., Hashim, M.M., Awad, H.F., Hu, B., Lei, S., Zeng, X. (2015) Quality assessment of pear juice under ultrasound and commercial pasteurization processing conditions. *Lebensmittel-Wissenschaft und-Technologie*, in press. DOI:10.1016/j.lwt.2015.05.005.

Wong, E., Vaillant, F., and Perez, A. (2010) Osmosonication of Blackberry Juice: Impact on Selected Pathogens, Spoilage Microorganisms, and Main Quality Parameters. *Journal of Food Science*, 75(7), M468–74. DOI:10.1111/j.1750-3841.2010.01730.x.

Yuan, Y., Hu, Y., Yue, T., Chen, T. and Lo, Y.M. (2009) Effect of ultrasonic treatments on thermoacidophilic *Alicyclobacillus acidoterrestris* in apple juice. *Journal of Food Processing and Preservation,* 33, 370–383. DOI:10.1111/j.1745-4549.2009.00407.x.

In: Applications of Ultrasound in the Beverage… ISBN: 978-1-63485-069-8
Editor: Juan Francisco García Martín

Chapter 3

ULTRASOUND-ASSISTED EXTRACTION OF BIOACTIVE COMPOUNDS FROM BYPRODUCTS OF FRUIT BEVERAGE INDUSTRY

Yang Tao, Yong Li, Ying Zhao, Fengzhi Lv and Yongbin Han*

College of Food Science and Technology,
Nanjing Agricultural University, Nanjing, China

ABSTRACT

Fruit beverage is closely associated with our daily life. However, fruit beverage industry produces million tons of byproducts every year, which is a considerable disposal problem. Byproducts from fruit beverage processing contain abundant bioactive compounds, including phenolics, carotenoids and seed oil. These bioactive compounds recovered from fruit beverage byproducts can be potentially used in food and pharmaceutical industries. As a novel and green food processing technology, ultrasound is an effective method to extract bioactive compounds from fruit beverage derived byproducts. Various factors can affect the efficiency of ultrasound during extraction, which can be divided into ultrasonic and non-ultrasonic factors. Furthermore, mathematical modeling is an

* Corresponding Author Email: hanyongbin@njau.edu.cn.

important tool to control the extraction process and scale up the ultrasound equipment. Both physical and empirical models can be considered for simulation. In general, ultrasound is a promising technology for the extraction of bioactive compounds from byproducts of fruit beverage industry.

Keywords: ultrasound, extraction, byproduct, fruit beverage, cavitation, modeling

INTRODUCTION

Fruit beverages play an important role in our daily life due to its easily digestible, thirst quenching, health-beneficial and other properties. There are many types of fruit beverages available in the market. Among them, alcoholic beverages produced from various fruits and fruit juices are the major fruit beverages. Meanwhile, fruit juices are usually blended with either soymilk or milk to improve the nutritional and sensory characteristics of final products (Rodríguez-Roque et al., 2015). Furthermore, several bioactive compounds, such as plant sterol, pyridoxine, arginine, selenium, etc., can be added into fruit juices to produce functional fruit beverages (Alemany-Costa et al., 2012; Gunathilake et al., 2013). In the last few years, fruit juice-based beverages fermented with probiotic bacteria and kefir microorganisms have also been developed to widen the choice of fruit beverages (Randazzo et al., 2016; Shori, 2016).

Fruit beverage industry produces a large amount of byproducts (around 20-60% of raw materials) every year as a result of beverage processing, such as seeds, peels, unusable pulp, wine lees from winemaking industry, etc. (Tao et al., 2014a; Amaya-Cruz et al. 2015). For example, the industrial production of fruit beverages generates about 15 million tons of citrus wastes and 10 million tons of grape wastes annually (Kammerer et al., 2004; Sanz-Puig et al., 2016). Traditionally, byproducts from fruit beverage are discarded or incinerated, thus producing a negative influence on environment. Otherwise, the byproducts are utilized as compost and animal feed. In the last decade, researchers are getting more and more concerned about developing new uses of these byproducts, such as production of colorants, dietary fiber, nutraceuticals and dietary supplements (Georgiev et al., 2014; Casquete et al., 2015).

Byproducts from fruit beverage industry are good sources of various bioactive compounds, including flavonoids, carotenoids, anthocyanins, vitamins, polysaccharides and many other phytochemicals. Thousands of papers report that compounds from fruit-based byproducts have many health-beneficial effects, such as antioxidant activity, antimicrobial activity, cytotoxic effect, anti-aging activity, anti-fatigue activity, antidiabetic activity, antityrosinase activity, anti-cancer activity, anti-inflammatory activity, antimutagenic activity, modification of neurodegenerative disorders, etc. (Aruomaet al., 2012; Chang et al., 2016). The exact bioactivities of compounds from fruit beverage related byproducts depend on the origin of byproducts. Generally, the recovery of bioactive compounds from byproducts is an important procedure to produce high-value co-products using residues from fruit beverage industry. Up to now, many effects have been made by researchers to extract bioactive compounds from fruit beverage derived byproducts effectively and economically.

The traditional method for extraction is solvent extraction, which means immersing the materials in various solvents with or without mechanical agitation and collecting the extracts after certain time (Chan et al., 2013). In the meantime, several physical treatments can be incorporated into solvent extraction process to enhance the mass transfer process during extraction, such as ultrasound irradiation, microwave irradiation, electric field treatment, high hydrostatic pressure processing, etc. Compared to conventional solvent extraction, the advantages of using ultrasound to assist extraction process can be summarized as high extraction efficiency, high reproducibility, high yield, low solvent consumption, easy-operating, low cost and low pollution to environment (Tao and Sun, 2015). According to the literature, ultrasound has already been applied for the recovery of bioactive compounds from different byproducts derived from fruit beverage industry (Tao et al., 2014a; Tao et al., 2014b; Drosou et al., 2015; González-Centeno et al., 2015).

In this chapter, the current state of the art of applications of ultrasound for extraction of bioactive compounds from byproducts derived from fruit beverage industry was reviewed. The principles of ultrasound for extraction and the factors affecting ultrasound extraction efficiency were discussed. Furthermore, the methodologies about simulation of ultrasound-assisted extraction from fruit beverage related byproducts were also discussed. All the information can provide basic guidance about utilizing ultrasound for the recovery of bioactive compounds from fruit beverage related byproducts.

Basic Knowledge About Ultrasound and Principle of Ultrasound-Assisted Extraction

According to the applied frequency and energy, ultrasound can be divided into two categories, namely low frequency/high power (frequency between 20 to 100 kHz and power between 10-1000 W cm^{-1}) ultrasound and high frequency/low power (frequency of MHz range and power < 1 W cm^{-1}) ultrasound (Kadam et al., 2015). Among them, low frequency/high power ultrasound is widely used to enhance solvent extraction process.

Generally, ultrasound equipment is composed of three parts, including generator, transducer and delivery system. To be specific, electricity is first converted to desired alternating current by ultrasound generator, thus driving ultrasound transducer assembly. Next, alternating current is changed to mechanical vibrations by ultrasound transducer. Finally, the vibration is conveyed to ultrasound reactor by delivery system (Tao and Sun, 2015). Ultrasound transducer is regarded as the key component for ultrasound reactor. There are mainly two types of ultrasound equipment that can be used for extraction, namely bath system and probe system. For bath system, ultrasound transducers can be mounted either on the bottom or wall of the bath. Regarding probe system, it may contain single probe or multi-probes. Furthermore, many ultrasound extractors are equipped with a cooling system, so as to control solvent temperature during extraction. The common figures about ultrasound extractor used in the laboratory are shown in Figure 1(a-c).

The principle about ultrasound-assisted extraction is mainly the physical and chemical effects of ultrasound cavitation. Since ultrasound wave is a kind of mechanical wave, its propagation through a medium produces a series of compression and rarefaction. Once the amplitude of ultrasound wave exceeds a threshold, cavitation phenomenon can be generated in liquid medium. Ultrasound cavitation is defined as the creation, expansion and collapse of microbubbles (Mason et al., 1996). According to the dynamic life of cavitation bubbles, ultrasound cavitation is divided into two types, namely stable cavitation and transient cavitation. Stable cavitation bubbles can exist for many cycles of compression and rarefaction, whereas transient cavitation bubbles only exist for a very short period followed by violent collapse (Tiwari, 2015). The violent collapse can produce extremely high temperature (2000-10000 K) and high pressure (100-1000 MPa) at a minuscule scale (Tiwari, 2015). These resulting high temperature and pressure can promote the disruption of raw material surface, thus enhancing the solvent penetration and the mass transfer of target compounds from inner structure of raw material to outside. Furthermore, the extremely high temperature and pressure produced can lead to the formation of highly reactive radicals. Taking water as example, water molecules can be broken down into OH^- and H^+. The resulting radicals can damage the biological cells they contact with, thus facilitating the release of extracted compounds from raw materials (Kurokawa et al., 2016).

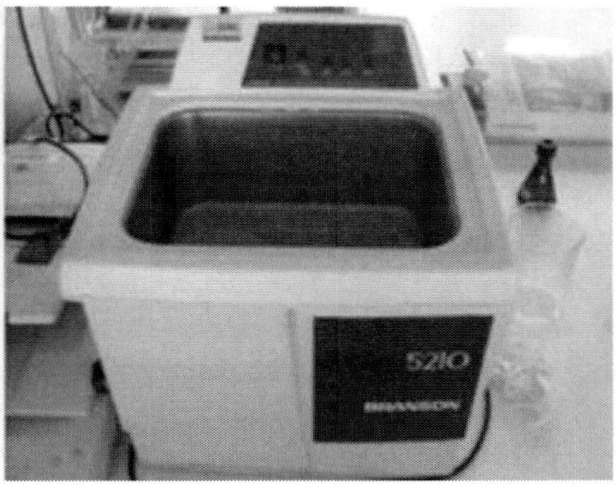

a b

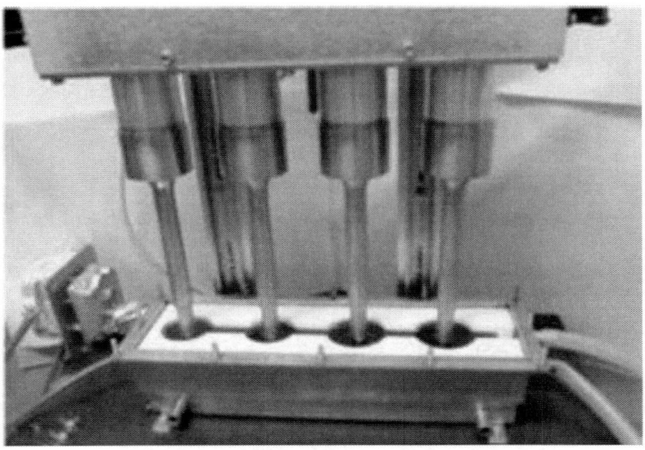

c

Figure 1. Images about common ultrasound extractors. a: bath system; b: single probe system; c: multi-probe system (Figure 1a and 1b from Chemat et al., 2011; Figure 1c from Alexandru et al., 2013).

On the other hand, strong eddies are produced during the vigorous vibration of cavitation bubbles. As a result, micro-streaming is formed due to the attraction of other small bubbles into ultrasonic fields (Hughes and Nyborg, 1962). Micro-streaming can provide a mechanical force to damage the raw material surface. Moreover, other physical phenomena produced by implosion of cavitation bubbles, including turbulence, agitation and inter-particle collision all contribute to the ultrasound enhancement of extraction process. Besides, ultrasound can promote the softening and swelling of raw materials in the solvent, which can also benefit the extraction process (Tiwari, 2015). The mechanism about ultrasound-assisted extraction is summarized in Figure 2. Although the aforementioned physical and chemical effects of ultrasound can benefit the extraction process, the question "which effect is the major contributor to extraction enhancement" has not yet been clarified clearly.

Extra attention should be paid to free radicals produced by ultrasound cavitation. Since the radicals have a very high chemical activity, they can not only destroy the biological issues, they may also react with extracted compounds. As a result, the extraction efficiency and the quality of final products are affected negatively. Ma et al. (2008) and Dong et al. (2010) reported that free radicals produced during sonication resulted in the degradation of phenolic acids and salvianolic acid B. Recently, Zhang et al. (2015) reported that free radicals generated by ultrasound can react with wine components, thus modifying wine physicochemical properties. The influence of free radicals on stability of extracted compounds should be taken into consideration in practical applications of ultrasound for extraction.

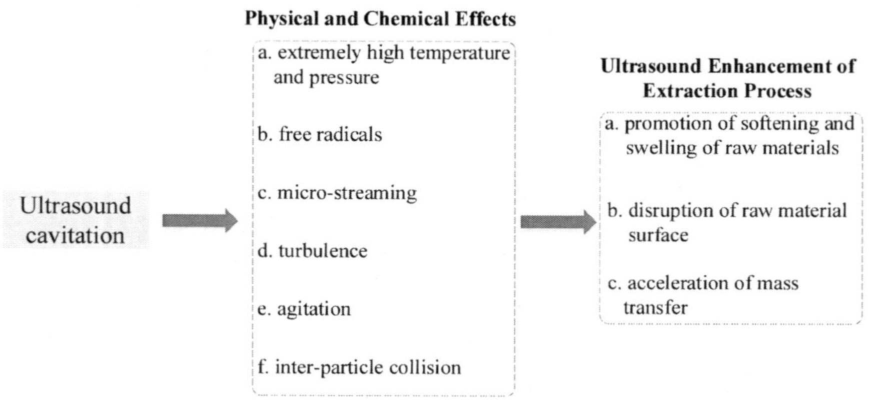

Figure 2. Mechanism about ultrasound-assisted extraction process.

Advance in Ultrasound-Assisted Extraction of Bioactive Compounds from Byproducts of Fruit Beverage Industry

According to the literature, most of bioactive compounds extracted by ultrasound from fruit beverage derived byproducts are phenolic compounds and their derivatives, carotenoids and seed oil. Generally, researchers concern about optimization of ultrasound-assisted extraction process, identification of bioactive compounds extracted from byproducts produced by fruit beverage industry and evaluation of bioactivity of resulting extracts.

Phenolic Compounds

Phenolic compounds are secondary metabolites in plants, which are derivatives of phosphate, phenylpropanoid, pentose and shikimate pathways (Balasundram et al., 2006). Phenolic compounds have exhibited a wide range of biological properties, such as antioxidant, antimicrobial, anti-inflammatory, anti-cancer, anti-allergenic and vasodilatory effects. Structurally, phenolics from fruits are composed of an aromatic ring with one or more hydroxyl substitutes. Phenolic compounds are a huge family, including both simple phenolic molecules and highly polymerized compounds (Bravo, 1998). According to their structures, phenolic compounds can be divided into the following classes, including simple phenolics and benzoquinones; hydroxybenzoic acids; acethophenones and phenylacetic acids; hydroxycinnamic acids and phenylpropanoids; napthoquinones; xanthones; stilbenes and anthraquinones; flavonoids and isoflavonoids; lignans and neolignans; bioflavonoids; lignins; and condensed tannins (Balasundram et al., 2006). Among them, phenolic acids, flavonoids and tannins are the primary dietary phenolic compounds, which are also the target compounds usually recovered from fruit beverage derived byproducts.

Compared to flavonoids and tannins, the structures of phenolic acids are relatively simple. Gallic, vanillic, syringic, **p**-hydroxybenzoic, caffeic, ferulic, and sinapic acids are common phenolic acids existing in fruits (Balasundram et al., 2006; Rodríguez-Roque et al., 2015; Heleno et al., 2015). As for flavonoids, they make up of the largest group of fruit phenolics, including flavones, flavonols, flavanols (or catechins), flavanones, flavanonols, isoflavones and anthocyanidins. The generic structures of major groups of flavonoids are shown in Figure 3 (Balasundram et al., 2006). Furthermore, the molecular weight of tannins is the highest among phenolic family in fruits. Tannins are divided into two categories, namely hydrolysable and condensed tannins. For details about phenolics from fruits, readers are suggested to read the literature focused on chemistry and biological activity of phenolics.

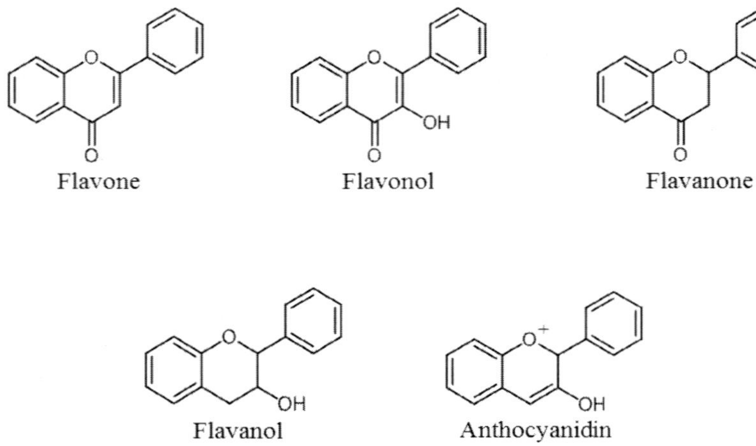

Figure 3. Generic structures of major groups of flavonoids (Balasundram et al., 2006).

Several papers about extraction of phenolic compounds from fruit beverage derived byproducts by means of ultrasound are reviewed and the main findings are summarized below. González-Centeno et al. (2015) used ultrasound to extract phenolic compounds from grape pomace and water was used as the extraction solvent. The results revealed that the ultrasound-assisted extraction time at 20, 35, and 50°C was 3, 4 and 8 times lower than conventional extraction time to obtain extracts with similar properties. Porto et al. (2013) extracted polyphenols from grape seeds using ultrasound. The authors found that the effectiveness of ultrasound to recover phenolics from grape seeds depended on the de-oiling method of grape seeds. Ultrasound extraction kinetics reached the equilibrium within 15 min when samples were defatted by Soxhlet. Tao et al. (2014b) investigated and modeled the extraction kinetics of phenolics from grape marc under sonication. It was found that temperature had a more profound influence than ultrasound energy on extraction yield. The effects of acoustic frequency and power density on the aqueous ultrasound-assisted extraction of phenolics from grape pomace were studied by González-Centeno et al. (2013). Both factors had significant influences on the extraction yield. In the study of Medina-Meza et al. (2015), anthocyanins and flavonols were extracted by ultrasound from both plum and grape peels. The extraction capacity of ultrasound was compared with that of pulsed electric fields. In plum peels, ultrasound behaved better than pulsed electric fields to extract anthocyanins and flavonols, whereas pulsed electric fields were the better choice for extraction from grape peels.

On the other hand, enzyme can be added to solvent during ultrasound-assisted extraction, since enzyme can catalyze the degradation of fruit cell wall effectively and accelerate the release of bioactive compounds. For example, Wu et al. (2015) established a simultaneous ultrasound/microwave enzymatic extraction method to recover phenolics from *Nitraria tangutorun* Bobr. juice by-products. The resulting extraction yield was 27.62-190.23% higher than that of conventional extraction methods. In summary, ultrasound has been demonstrated as a promising tool for the recovery of phenolic compounds from fruit beverage derived byproducts.

Carotenoids

Carotenoids are lipid-soluble pigments that can be divided into two classes, namely xanthophylls and carotenes. Xanthophylls are molecules containing oxygen, such as zeaxanthin and lutein, while carotenes are non-oxygenated molecules, including lycopene and α-carotene (Strati and Oreopoulou, 2014). The generic carotenoid structure is shown in Figure 4. Generally, fruits can provide most of α-carotene and β-carotene and orange fruits are good sources of α-cryptoxanthin (Amorim-Carrilho et al., 2014).

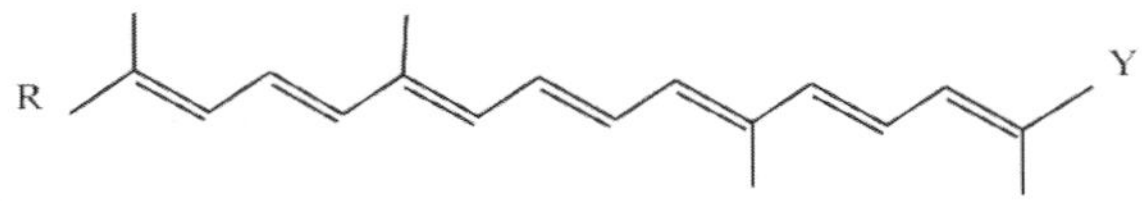

Figure 4. Generic structures of major groups of carotenoids (Balasundram et al., 2006).

Regarding the biological activities of carotenoids, almost all the carotenoids with both high and low degrees have the antioxidant capacity. Some of carotenoids are pro-vitamin A carotenoids, which can be changed to vitamin A. Vitamin A has several biological activities, including prevention of eye diseases, scaly skin, rough, bone development and retarded tooth (Jaswir et al., 2011). In the meantime, carotenoid consumption has therapeutic effects on heart and vascular diseases, age-related macular degeneration, cataracts, cancer and degenerative diseases (Amorim-Carrilho et al., 2014). Furthermore, there are some studies reporting that carotenoids have shown certain potential to modulate gene expression (Hung et al., 2008). Due to the aforementioned bioactive activity of carotenoids, carotenoids are a group of important bioactive compounds extracted from fruit beverage derived byproducts.

Carotenoids in peach palm fruit byproducts were extracted by ultrasound in the study of Ordóñez-Santos et al. (2015). Sunflower oil was used as the solvent and the effects of ultrasound intensity, temperature, and extraction time on extraction yield were investigated. The optimum extraction conditions were identified to be 1528 W/cm^2 for ultrasound intensity, 35°C for extraction temperature and 30 min for extraction time. Similarly, in the study of Sun et al. (2011) about ultrasound-assisted extraction from citrus peels, the influences of several extrinsic and intrinsic factors on carotenoid yield were studied. According to the results, particle size was the key factor affecting extraction efficiency. Meanwhile, ultrasound intensity had a complicated influence on extraction yield. Furthermore, lycopene in tomato processing wastes was also extracted by ultrasound in the study of Kumcuoglu et al. (2014). The mixture of hexane, acetone and ethanol at a ratio of 2:1:1 (v:v:v) with 0.05% (butylated hydroxytoluene, w:v) was used as the solvent. The highest extraction yield was obtained when solvent-to-solid ratio was 35:1 (v:w), ultrasound power was 90 W and extraction time was 30 min. In summary, all the published studies prove that ultrasound is more efficient than conventional extraction method for the recovery of carotenoids from byproducts produced by fruit beverage industry.

Seed Oil

Fruit seeds are the biomass waste in the fruit beverage industry. As a matter of fact, seeds constitute a considerable portion of fruit weight. On the other hand, the seeds of many fruits, such as grape, blueberry, raspberry, papaya, pomegranate, etc., are rich in vitamin E (tocotrienols and tocopherols) and unsaturated fatty acids (Duba and Fiori, 2015). Fruit seed oil has a high antioxidant activity, thus making it attractive in medical, pharmaceutical and cosmetic applications. Papaya seed oil was extracted ultrasonically by Samaram et al. (2014). In this study, the quality of extracted seed oil was evaluated from five aspects, including crystallization and thermal behavior, color, saturation degree and oxidative stability. The results revealed that ultrasound extracted seed oil had a high quality with low amount of unsaponifiable compounds. In another paper of Samaram et al. (2015), the ultrasound-assisted extraction process was optimized. The resulting optimum conditions for extraction of seed oil from papaya seed were 62.5°C for extraction temperature, 700 W for ultrasound power and 38.5 min for extraction time. In the study of Goula (2013), the author investigated the extraction kinetics of pomegranate seed oil under sonication and optimized the extraction conditions. The optimum conditions were found to be extraction temperature, 20°C; amplitude level, 60%; solvent-to-solid ratio, 20/1; seed particle size, 0.2 mm; pulse duration/pulse interval ratio, 5/15. Furthermore, Porto et al. (2013) compared the extraction yield and fatty oil composition of grape seed oil extracted by ultrasound and Soxhlet extraction. The results showed that grape seed oil yield obtained after 20-kHz sonication at 150 W for 30 min was comparable to that obtained after 6-h Soxhlet extraction. At the same time, the major composition of fatty acids in grape seed oil obtained using both extraction methods had no significant differences. As regarded to other reports about ultrasound-assisted extraction of seed oil, many of them are concerned about the optimization of extraction process and analysis of product composition.

Factors Affecting Extraction Efficiency During Ultrasound-Assisted Extraction of Bioactive Compounds from Fruit Beverage Derived Byproducts

In order to fulfill the potential of ultrasound to extract bioactive compounds from fruit beverage derived byproducts, the factors affecting extraction efficiency should be taken into account. These factors can be divided into two categories, including ultrasonic factors and non-ultrasonic factors (Tao and Sun, 2015).

Ultrasonic Factors

Ultrasonic factors include ultrasound energy, ultrasound duration, ultrasound frequency and ultrasound treatment mode.

Ultrasound energy is expressed as ultrasound power, ultrasound intensity, acoustic energy density, and ultrasound amplitude in different papers.

Ultrasound amplitude is proportional to ultrasound power, intensity and acoustic energy density. It should be pointed out that amplitude only represents the energy required to drive ultrasound transducer, instead of the absolute energy entering extraction system (Tiwari, 2015). To obtain the value of ultrasound energy in extractor, either calorimetry or chemical dosimetry method can be used. Theoretically, the acoustic energy can markedly affect the acoustic amplitude and the dynamics of cavitation bubbles (Merouani et al., 2013), and subsequently influence the extraction progress. In a proper range, the initial extraction rate and equilibrium extraction yield increase with the increase of ultrasound energy, since high energy can benefit the disruption of cell structure and facilitate the release of target compounds from raw materials. The typical extraction curves of bioactive compounds from fruit beverage derived byproducts under sonication at different acoustic energy densities are shown in Figure 5 (Tao et al., 2014b). However, it should be pointed out that high amplitude level may inhibit the formation of ultrasound cavitation, cause erosion of ultrasound probe and decrease the life of ultrasound equipment, while high ultrasound intensity may result in the degradation of target compounds (Tiwari, 2015).

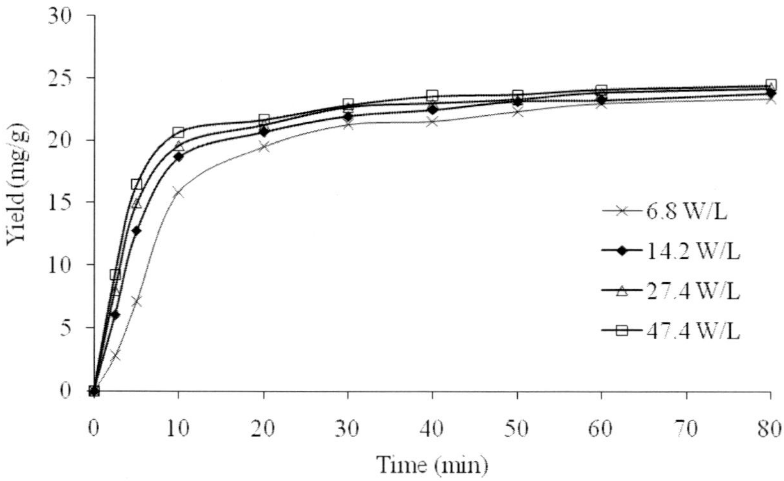

Figure 5. Extraction kinetic curves of phenolic compounds from grape marc under sonication at different acoustic energy densities (Tao et al., 2014b).

As for ultrasound frequency, ultrasound frequency used for extraction usually ranges from 20 to 100 kHz. Several studies reported that low ultrasound frequency (20-40 kHz) can promote the disruption of cell structure, thus maximizing the extraction yield (Cravotto et al., 2008; Shirsath et al., 2012; González-Centeno et al., 2013). Different results were reported by Vetal et al. (2013), who found that the extraction yield of ursolic acid from basil during sonication at 40 kHz was significantly higher than that obtained during sonication at 25 kHz. Furthermore, the effect of ultrasound frequency on extraction yield also depends on the input ultrasound energy (González-Centeno et al., 2013). High ultrasound energy is required at high frequencies to conquer the cohesive forces between solvent and raw materials (Niazi et al., 2014). From a physical point of view, ultrasound frequency is responsible for the physical and chemical effects of ultrasound. At low frequencies (20-100 kHz), the physical effects of ultrasound are dominant, while at frequencies between 200-500 kHz, chemical effects of ultrasound are more dominant (Tiwari, 2015).

In most cases, the extraction kinetic curve of ultrasound-assisted extraction is similar to that of traditional solvent extraction, which is characterized by a rapid increase in extraction yield at beginning followed by steady increase until reaching the extraction equilibrium. The rapid increase in extraction yield corresponds to the washing stage, while the slow increase in extraction yield refers to the diffusion stage (Tao et al., 2014b). Generally, ultrasound duration should not exceed the time needed to achieve the extraction equilibrium. Considering that it may be time-consuming for ultrasound-assisted extraction to reach the equilibrium, it is suggested by the authors that the ultrasound extraction process could be ended before reaching the equilibrium point, thus saving the time and cost for extraction. The proper ultrasound duration required for recovering bioactive compounds from fruit beverage derived byproducts can be determined through optimizing the extraction process.

The last ultrasonic factor is ultrasound treatment mode. Ultrasound-assisted extraction can be performed either in a continuous or pulsed mode in practical application. Compared to continuous mode, pulsed mode is regarded as an energy saving mode (You et al., 2014; Tao and Sun, 2015). If ultrasound duration is long or ultrasound energy is high during extraction, pulsed mode is recommended to avoid rapid increase in solvent temperature due to thermal effect of ultrasound and minimize the negative effect of ultrasound on the stability of target compounds.

Non-Ultrasonic Factors

Non-ultrasonic factors include solvent type, particle size, extraction temperature, and solvent-to-solid ratio.

Solvent type is an important non-ultrasonic factor. The physical properties of solvent, including polarity, surface tension, viscosity, density, vapor pressure and diffusivity should be considered when it comes to select solvent used for extraction. These properties are associated with solvent's molecular affinity with target compounds and intensity of ultrasound cavitation in solvent. On one hand, to produce high-intensity ultrasound cavitation, solvents with low vapor pressure, low surface tension and low viscosity should be used (Tiwari, 2015). On the other hand, the boiling point of solvent should be low, thus making the removal of solvent from target compounds easily (Tao and Sun, 2015). The common solvents used for extraction of phenolics from fruit beverage derived byproducts include ethanol, methanol and their mixtures with water (Vikhu et al., 2008; Tao et al., 2014b; Rodrigues et al., 2015). Meanwhile, the common solvents used for the recovery of carotenoids from fruit beverage derived byproducts are ethanol, hexane, methanol and their combinations (Macías-Sánchez et al., 2009; Luengo et al., 2014; Zaghdoudi et al., 2015). Furthermore, *n*-hexane is widely used as solvent for the ultrasound-assisted extraction of seed oil.

As for particle size of raw materials, particles with small size have a large surface area, thus making the penetration of ultrasound waves into raw materials easier. However, particle size should not be too small. Otherwise, particles may float on the surface of solvent, which can weaken the ultrasound effect (Khan et al., 2010).

Temperature plays a profound effect on the effectiveness of ultrasound during sonication. In a proper range, extraction yield increases with the increase of temperature during sonication, since a high temperature can accelerate the swelling and softening of raw materials and increase the solubility of target compounds (Tao et al., 2014a; Tao et al., 2014b). However, temperature during sonication should not be too high, since target compounds may be degraded at high temperatures (Gouda, 2013). Meanwhile, temperature rising can lead to the decrease of surface tension and increase of vapor pressure of solvent, thus decreasing the energies released from the collapse of ultrasound cavitation bubbles (Zhang et al., 2008). Furthermore, high temperatures mean high energy consumption.

During ultrasound-assisted extraction, a high solvent-to-solid ratio can benefit extraction process due to a high concentration gradient between bulky solvent and raw materials, which is the driving force during mass transfer (Tao and Sun, 2015). Nevertheless, an extremely high level of solvent-to-solid ratio should be avoided due to the high consumption of solvent and the increasing cost for concentration and filtration of extracts in the following steps (Cissé et al., 2011).

Simulation of Ultrasound-Assisted Extraction of Bioactive Compounds from Fruit Beverage Derived Byproducts

From an engineering point of view, it is important to model the ultrasound-assisted extraction process, so as to provide guidance about controlling the extraction process automatically and scaling up the sono-exactor.

Table 1 summarizes some models that can be used to simulate the extraction process. All these models are empirical, semi-empirical and theoretical models. Among all these models, responsible surface model has been widely used to investigate the effects of different experimental factors on extraction yield and optimize the extraction process. However, responsible surface model only considers the polynomial relationship between experimental factors and extraction yield. On the other hand, artificial neural network is regarded as a better strategy for extraction modeling due to its capacity to correlate any forms of non-linear relationship (Ghoreishi and Heidari, 2013; Tao et al., 2014a).

Response surface model, artificial neural network and many empirical models do not account for the phenomena governing the mass transfer process during sonication. Instead, the diffusion model based on the Fick's second law can describe the diffusion of target compounds within raw materials, thus containing physical significance. However, to use diffusion model for modeling, several assumptions should be made, such as constant effective diffusion coefficient of target compounds, no external mass transfer resistance, etc. Diffusion model can be solved either analytically or numerically (Tao et al., 2014b; Tao et al., 2014c). To get the analytical solution of diffusion model, the geometry of raw materials needs to be specified to slab, cylindrical or spherical shapes.

Table 1. Several mathematical models suitable for simulating ultrasound-assisted extraction process

Model	Model information	Reference
Responsible surface model	$Y=\beta_0+\sum_{i=1}\beta_i X_i+\sum_{i=1}\beta_{ii}X_i^2+\sum_{i=1}\sum_{j=i+1}\beta_{ij}X_iX_j$ where X_i and X_j are experimental factors, Y is extraction yield, β_0, β_i, β_{ii} and β_{ij} are the constant, linear, quadratic and cross product coefficients, respectively.	Tao and Sun, 2015; Tomšik et al., 2016
Artificial neural network	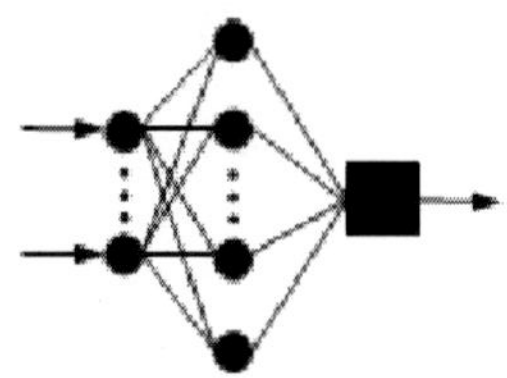 The neurons in input layer represent the experimental factors and the neuron in output layer denotes the extraction yield. The performance of neural network for prediction can be optimized through training.	Tao et al., 2014a; Dahmoune et al., 2015
Diffusion model based on the Fick's second law	$\frac{\partial C(x,t)}{\partial t}=D_e\frac{1}{x^{v-1}}\frac{\partial}{\partial x}\left[x^{v-1}\frac{\partial C(x,t)}{\partial x}\right]$ where C is the concentration of target compounds in raw materials, D_e is effective diffusion coefficient, t is time, x is radial distance in the diffusion direction, v is shape index.	Tao et al., 2014b
General second-order model	$C(t)=\frac{C_e^2kt}{1+C_ekt}=\frac{C_et}{1/C_ek+t}$ where C is the concentration of target compounds in solvent, C_e is equilibrium concentration in solvent, k is second-order extraction rate constant, t is time.	Goula, 2013
Two site kinetic model	$\frac{M_t}{M_\infty}=1-Fe^{-k_1t}-(1-F)e^{-k_2t}$	Tao et al., 2014c

Model	Model information	Reference
	where M_t is the amount of target compounds removed from raw materials at time t, M_∞ is the maximum amount of target compounds extracted after infinite time. F is the portion of target compounds extracted rapidly, (1-F) is the portion of target compounds extracted slowly, k_1 is the first-order rate constant representing the rapid extracted portion, k_2 is the first-order rate constant representing the slow extracted portion	
Parabolic diffusion model	$C = A_0 + A_1 t^{1/2}$ where C is extraction yield, A_0 is washing coefficient, A_1 is diffusion rate constant.	Kitanović et al., 2008
Power law model	$C = Bt^n$ where C is extraction yield, B is the parameter incorporating the characteristics of the extraction system.	Kitanović et al., 2008
Hyperbolic model (Peleg's model)	$C = \frac{C_1 t}{1 + C_2 t}$ where C is extraction yield, C_1 is extraction rate at the very beginning, C_2 is constant related to maximum extraction yield, t is time	Kitanović et al., 2008; Tiwari, 2015
Weibull model	$C = 1 - \exp(-\frac{t^m}{D})$ where C is extraction yield, D is scale parameter, m is shape parameter, t is time.	Kitanović et al., 2008
Elovich's model	$C = E_0 + E_1 \ln t$ where C is extraction yield, E_0 and E_1 are fitting parameters.	Kitanović et al., 2008
Combined second-order-diffusional kinetic model	$c = \frac{c_e^w t}{T_{w(1/2)} + t} + c_e^d (1 - \exp(-k_d t))$ where c is concentration of target compounds in solvent, c_e^w is equilibrium concentration of target compounds in solvent due to washing mechanism, c_e^d is equilibrium concentration of target compounds due to diffusion mechanism, k_d is diffusional mechanism kinetics constant, $T_{w(1/2)}$ is half-life time for the second-order model in diffusion mechanism, t is time.	Linares et al., 2010

Note: the units of all the parameters in these equations are decided depending on each specific extraction process.

Furthermore, none of the aforementioned models consider the ultrasound cavitation phenomenon. It is a tough task to model cavitation during ultrasound-assisted extraction and quantify the effect of ultrasound cavitation on extraction process. Kim et al. (2009) used the acoustic model and heat transfer model of COMMSOL Multiphysics to predict the pressure and temperature profiles during sonication, thus establishing the relationship between extraction yield and cavitation density. Numerical modeling using commercial software, such as COMMSOL Multiphysics, can provide a novel strategy for simulation of ultrasound-assisted extraction process. Complicated models can be built with the assistance of commercial software. Researchers are encouraged to model the extraction process by means of numerical methods in future studies.

CONCLUSION

This chapter is focused on ultrasound-assisted extraction of bioactive compounds from byproducts of fruit beverage industry. Fruit beverage industry produces a large amount of byproducts each year. Fruit beverage derived byproducts are rich in many bioactive compounds, such as phenolics, carotenoids, seed oil, etc. The recovery of bioactive compounds from the byproducts can not only reduce their negative impacts on environment, but also produce some economic benefits. Ultrasound in the frequency between 20-100 kHz is a powerful technology to extract bioactive compounds from byproducts of fruit beverage industry. There are many papers about extraction of bioactive compounds using ultrasound and this technology has also been applied on fruit beverage derived byproducts. The enhancement of extraction process is mainly due to the ultrasound cavitation phenomenon. To fulfill the potential of ultrasound during extraction, the influences of ultrasound energy, ultrasound frequency, ultrasound duration, processing mode, particle size, temperature, solvent type and solvent-to-solid ratio should be taken into consideration. Mathematical modeling is very important for controlling the ultrasound-assisted extraction process and scaling up the sono-extractors. Although many models have been developed, the models including the information about ultrasound cavitation is deficient. With the assistance of advanced commercial software, numerical modeling of ultrasound cavitation during extraction can be easier. Future studies should pay more attention to numerical modeling and scale-up of sono-extractor. Lastly, more researches are also needed to develop novel products based on bioactive compounds extracted from fruit beverage byproducts, so as to meet different demands of consumers.

REFERENCES

Alemany-Costa, L., González-Larena, M., García-Llatas, G., Alegría, A., Barberá, R., Sánchez-Siles, L.M., Lagarda, M.J., (2012). Sterol stability in functional fruit beverages enriched with different plant sterol sources. *Food Res. Int.* 48, 265-270.

Alexandru, L., Cravotto, G., Giordana, L., Binello, A., Chemat, F., (2013). Ultrasound-assisted extraction of clove buds using batch- and flow-reactors: A comparative study on a pilot scale. *Innov. Food Sci. Emerg. Technol.* 20, 167-172.

Amaya-Cruz, D.M., Rodríguez-González, S., Pérez-Ramírez, I.F., Loarca-Piña, G., Amaya-Llano, S., Gallegos-Corona, M.A., Reynoso-Camacho, R., (2015). Juice by-products as a source of dietary fibre and antioxidants and their effect on hepatic steatosis. *J. Funct. Food.* 17, 93-102.

Amorim-Carrilho, K.T., Cepeda, A., Fente, C., Regal, P., (2014). Review of methods for analysis of carotenoids. *Trac-Trends Anal. Chem.* 56, 49-73.

Aruoma, O., Coles, L.S., Landes, B., Repine, J.E., (2012). Functional benefits of ergothioneine and fruit- and vegetable-derived nutraceuticals: Overview of the supplemental issue contents. *Prev. Med.* 54, S4-S8.

Balasundram, N., Sundram, K., Samman, S., (2006). Phenolic compounds in plants and agri-industrial by-products: Antioxidant activity, occurrence, and potential uses. *Food Chem.* 99, 191-203.

Bravo, L., (1998). Polyphenols: chemistry, dietary sources, metabolism, and nutritional significance. *Nutr. Rev.* 56, 317-333.

Casquete, R., Castro, S.M., Martín, A., Ruiz-Moyano, S., Saraiva, J.A., Córdoba, M.G., Teixeira, P., (2015). Evaluation of the effect of high pressure on total phenolic content, antioxidant and antimicrobial activity of citrus peels. *Innov. Food Sci. Emerg. Technol.* 31, 37-44.

Chan, C.-H., Yusoff, R., Ngoh, G.-C., (2013). Modeling and kinetics study of conventional and assisted batch solvent extraction. *Chem. Eng. Res. Des.*, 92, 1169-1186.

Chang, S.K., Alasalvar, C., Shahidi, F., (2016). Review of dried fruits: Phytochemicals, antioxidant efficacies, and health benefits. *J. Funct. Food.* 21, 113-132.

Chemat, F., Huma, Z., Khan, M.K., (2011). Applications of ultrasound in food technology: Processing, preservation and extraction. *Ultrason. Sonochem.* 18, 813-835.

Cravotto, G., Boffa, L., Mantegna, S., Perego, P., Avogadro, M., Cintas, P., (2008). Improved extraction of vegetable oils under high-intensity ultrasound and/or microwaves. *Ultrason. Sonochem.* 15, 898-902.

Cissé, M., Bohuon, P., Sambe, F., Kane, C., Sakho, M., Dornier, M., (2011). Aqueous extraction of anthocyanins from *Hibiscus sabdariffa*: Experimental kinetics and modeling. *J. Food Eng.* 109, 16-21.

Dahmoune, F., Remini, H., Dairi, S., Aoun, O., Moussi, K., Bouaoudia-Madi, N., Adjeroud, N., Kadri, N., Lefsih, K., Boughani, L., Mouni, L., Nayak, B., Madani, K., (2015). Ultrasound assisted extraction of phenolic compounds from *P. lentiscus* L. leaves: Comparative study of artificial neural network (ANN) versus degree of experiment for prediction ability of phenolic compounds recovery. *Ind. Crop. Prod.* 77, 251-261.

Dong, J., Liu, Y., Liang, Z., Wang, W., (2010). Investigation on ultrasound-assisted extraction of salvianolic acid B from *Salvia miltiorrhiza* root. *Ultrason. Sonochem.* 17, 61-65.

Drosou, C., Kyriakopoulou, K., Bimpilas, A., Tsimogiannis, D., Krokida, M., (2015). A comparative study on different extraction techniques to recover red grape pomace polyphenols from vinification byproducts. *Ind. Crop. Prod.* 75, 141-149.

Duba, K.S., Fiori, L., (2015). Supercritical CO_2 extraction of grape seed oil: Effect of process parameters on the extraction kinetics. *J. Supercrit. Fluids.* 98, 33-43.

Georgiev, V., Ananga, A., Tsolova, V., (2014). Recent advances and uses of grape flavonoids as nutraceuticals. *Nutrients*, 6, 391-415.

Ghoreishi, S.M., Heidari, E., (2013). Extraction of Epigallocatechin-3-gallate from green tea via supercritical fluid technology: Neural network modeling and response surface optimization. *J. Supercrit. Fluids*, 74, 128-136.

González-Centeno, M.R., Knoerzer, K., Sabarez, H., Simal, S., Rosselló, C., Femenia, A., (2013). Effect of acoustic frequency and power density on the aqueous ultrasonic-assisted extraction of grape pomace (*Vitis vinifera* L.) - A response surface approach. *Ultrason. Sonochem.* 21, 2176-2184.

González-Centeno, M.R., Comas-Serra, F., Femenia, A., Rosselló, C., Simal, S. (2015). Effect of power ultrasound application on aqueous extraction of phenolic compounds and antioxidant capacity from grape pomace (*Vitis vinifera* L.): Experimental kinetics and modeling. *Ultrason. Sonochem.* 22, 506-514.

Goula, A.M. (2013). Ultrasound-assisted extraction of pomegranate seed oil - Kinetic modeling. *J. Food Eng.* 117, 492-498.

Gunathilake, K.D.P.P., Wang, Y., Rupasinghe, H.P.V., (2013). Hypocholesterolemic and hypotensive effects of a fruit-based functional beverage in spontaneously hypertensive rats fed with cholesterol-rich diet. *J. Funct. Food.* 5, 1392-1401.

Heleno, S.A., Martins, A., Queiroz, M.J.R.P., Ferreira, I.C.F.R., (2015). Bioactivity of phenolic acids: Metabolites versus parent compounds: A review. *Food Chem.* 501-513.

Hughes, D.E., Nyborg, W.L., (1962). Cell disruption by ultrasound. *Science*, 138, 108-114.

Hung, C., Huang, T., Chen, B., Shieh, J., Wu, P., Wu, W., (2008). Lycopene inhibits TNF-α-induced endothelial ICAM-1 expression and monocyte-endothelial adhesion. *Eur. J. Pharmacol.* 586, 275-282.

Jaswir, I., Noviendri, D., Reno, F.H., Octavianti, F. (2011). Carotenoids: sources, medicinal properties and their application in food and nutraceuticalindustry. *J. Med. Plants Res.* 5, 7119-7131.

Kadam, S.U., Tiwari, B.K., Álvarez, C., O'Donnell, C.P., (2015). Ultrasound applications for the extraction, identification and delivery of food proteins and bioactive peptides. *Trends Food Sci. Technol.* 46, 60-67.

Kammerer, D., Claus, A., Carle, R., Schieber, A., (2004). Polyphenol screening of pomace from red and white grape varieties (*Vitisvinifera* L.) by HPLC-DAD-MS/MS. *J. Agric. Food Chem.*52, 4360-4367.

Khan, M. K., Abert-Vian, M., Fabiano-Tixier, A., Dangles, O., Chemat, F., (2010). Ultrasound-assisted extraction of polyphenols (flavanone glycosides) from orange (*Citrus sinensis* L.) peel. *Food Chem.* 119, 851-858.

Kim, H.-J., Chi, M.-H., Hong, I.K., (2009). Effect of ultrasound irradiation on solvent extraction process. *J. Ind. Eng. Chem.* 15, 919-928.

Kitanović, S., Milenović, D., Veljković, V.B., (2008). Empirical kinetic models for the resinoid extraction from aerial parts of St. John's wort (*Hypericum perforatum* L.). *Biochem. Eng. J.* 41, 1-11.

Kumcuoglu, S., Yilmaz, T., Tavman, S., (2014). Ultrasound assisted extraction of lycopene from tomato processing wastes. *J. Food Sci. Technol.*51, 4102-4107.

Kurokawa, M., King, P.M., Wu, X., Joyce, E.M., Mason, T.J., Yamamoto, K., (2016). Effect of sonication frequency on the disruption of algae. *Ultrason. Sonochem.* 31, 157-163.

Linares, A.R., Hase, S.L., Vergara, M.L., Resnik, S.L., (2010). Modeling yerba mate aqueous extraction kinetics: Influence of temperature. *J. Food Eng.* 97, 471-477.

Luengo, E., Condón-Abanto, S., Condón, S., Álvarez, I., Raso, J., (2014). Improving the extraction of carotenoids from tomato waste by application of ultrasound under pressure. *Sep. Purif. Technol.*, 136, 130-136.

Ma, Y.Q., Ye, X.Q., Fang, Z.X., Chen, J.C., Xu, G.H., Liu, D.H., (2008). Phenolic compounds and antioxidant activity of extracts from ultrasonic treatment of Satsuma mandarin (*Citrus unshiu* Marc.) peels. *J. Agric. Food Chem.*, 56, 5682-5690.

Macías-Sánchez, M.D., Mantell, C., Rodríguez, M., Ossa, E.M., Lubián, L.M., Montero, O., (2009). Comparison of supercritical fluid and ultrasound-assisted extraction of carotenoids and chlorophyll a from *Dunaliella salina*. *Talanta*, 77, 948-952.

Mason, T.J., Paniwnyk, L., Lorimer, J.P., (1996). The use of ultrasound in food technology. *Ultrason. Sonochem.* 3, S253- S256.

Medina-Meza, I.G., Barbosa-Cánovas, G., (2015). Assisted extraction of bioactive compounds from plum and grape peels by ultrasonics and pulsed electric fields. *J. Food Eng.* 166, 268-275.

Merouani, S., Hamdaoui, O., Rezgui, Y., Guemini, M., (2013). Effects of ultrasound frequency and acoustic amplitude on the size of sonochemically active bubbles - Theoretical study. *Ultrason. Sonochem.* 20, 815-819.

Niazi, S., Hashemabadi, S.H., Noroozi, S., (2014). Numerical simulation of operational parameters and sonoreactor configurations for the highest possibility of acoustic cavitation in crude oil. *Chem. Eng. Commun.* 201, 1340-1359.

Ordóñez-Santos, L.E., Pinzón-Zarate, L.X., González-Salcedo, L.O., (2015). Optimization of ultrasonic-assisted extraction of total carotenoids from peach palm fruit (*Bactris gasipaes*) by-products with sunflower oil using response surface methodology. *Ultrason. Sonochem.* 27, 560-566.

Porto, C.D., Porretto, E., Decorti, D. (2013). Comparison of ultrasound-assisted extraction with conventional extraction methods of oil and polyphenols from grape (*Vitisvinifera* L.) seeds. *Ultrason. Sonochem.* 20, 1076-1080.

Randazzo, W., Corona, O., Guarcello, R., Francesca, N., Germanà, M.A., Erten, H., Moschetti, G., Settanni, L., (2016). Development of new non-dairy beverages from Mediterranean fruit juices fermented with water kefir microorganisms. *Food Microbiol.* 54, 40-51.

Rodrigues, S., Fernandes, F.A.N., Brito, E.S., Sousa, A.D., Narain, N., (2015). Ultrasound extraction of phenolics and anthocyanins from jabuticaba peel. *Ind. Crop. Prod.* 69, 400-407.

Rodríguez-Roque, M., Ancos, B., Sánchez-Moreno, C., Cano, M.P., Elez-Martínez, P., Martín-Belloso, O., (2015). Impact of food matrix and processing on the in vitro bioaccessibility of vitamin C, phenolic compounds, and hydrophilic antioxidant activity from fruit juice-based beverages. *J. Funct. Food.* 14, 33-43.

Samaram, S., Mirhosseini, H., Tan, C.P., Ghazali, H.M., (2014). Ultrasound-assisted extraction and solvent extraction of papaya seed oil: Crystallization and thermal behavior, saturation degree, color and oxidative stability. *Ind. Crop. Prod.* 52, 702-708.

Samaram, S., Mirhosseini, H., Tan, C.P., Ghazali, H.M., Bordbar, S., (2015). Optimisation of ultrasound-assisted extraction of oil from papaya seed by response surface methodology: Oil recovery, radical scavenging antioxidant activity, and oxidation stability. *Food Chem.* 172, 7-17.

Sanz-Puig, M., Pina-Pérez, M.C., Martínez-López, A., Rodrigo, D., (2016). *Escherichia coli* O157:H7 and *Salmonella* Typhimurium inactivation by the effect of mandarin, lemon, and orange by-products in reference medium and in oat-fruit juice mixed beverage. *LWT-Food Sci. Technol.* 66, 7-14.

Shirsath, S.R., Sonawane, S.H., Gogate, P.R.(2012). Intensification of extraction of natural products using ultrasonic irradiations. A review of current status. *Chem. Eng. Process.* 53, 10-23.

Shori, A.B., (2016). Influence of food matrix on the viability of probiotic bacteria: A review based on dairy and non-dairy beverages. *Food Bioscience*, 13, 1-8.

Strati, I.F., Oreopoulou, V., (2014). Recovery of carotenoids from tomato processing by-products -a review. *Food Res. Int.* 65, 311-321.

Sun, Y., Liu, D., Chen, J., Ye, X., Yu, D., (2011). Effects of different factors of ultrasound treatment on the extraction yield of the all-*trans*-β-carotene from citrus peels. *Ultrason. Sonochem.* 18, 243-249.

Tao, Y., Wu, D., Zhang, Q.-A., Sun, D.-W., (2014a). Ultrasound-assisted extraction of phenolics from wine lees: Modeling, optimization and stability of extracts during storage. *Ultrason. Sonochem.* 21. 706-715.

Tao, Y., Zhang, Z., Sun, D.-W. (2014b). Kinetic modeling of ultrasound-assisted extraction of phenolic compounds from grape marc: Influence of acoustic energy density and temperature. *Ultrason. Sonochem.* 21, 1461-1469.

Tao, Y., Zhang, Z., Sun, D.-W., (2014c). Experimental and modeling studies of ultrasound-assisted release of phenolics from oak chips into model wine. *Ultrason. Sonochem.* 21, 1839-1848.

Tao, Y., Sun, D.-W., (2015). Enhancement of food processes by ultrasound: a review. *Crit. Rev. Food Sci. Nutr.* 55, 570-594.

Tiwari, B.K. (2015). Ultrasound: A clean, green extraction technology. *Trac-Trends Anal. Chem.* 71, 100-109.

Tomšik, A., Pavlić, B., Vladić, J., Ramić, M., Brindza, J., Vidović, S., (2016). Optimization of ultrasound-assisted extraction of bioactive compounds from wild garlic (*Allium ursinum* L.). *Ultrason. Sonochem.* 29, 502-511.

Vetal, M.D., Lade, V.G., Rathod, V.K., (2013). Extraction of ursolic acid from Ocimum sanctum by ultrasound: process intensification and kinetic studies. *Chem. Eng. Process.* 69, 24-30.

Vilkhu, K., Mawson, R., Simons, L., Bates, D., (2008). Applications and opportunities for ultrasound assisted extraction in the food industry - A review. *Innov. Food Sci. Emerg. Technol.* 9, 161-169.

Wu, D., Gao, T., Yang, H., Du, Y., Li, C., Wei, L., Zhou, T., Lu, J., Bi, H., (2015). Simultaneous microwave/ultrasonic-assisted enzymatic extraction of antioxidant ingredients from *Nitraria tangutorun* Bobr. juice by-products. *Ind. Crop. Prod.* 66, 229-238.

You, Q., Yin, X., Ji, C., (2014). Pulsed counter-current ultrasound-assisted extraction and characterization of polysaccharides from *Boletus edulis*. *Carbohydr. Polym.*, 101, 379-385.

Zaghdoudi, K., Pontvianne, S., Framboisier, X., Achard, M., Kudaibergenova, R., Ayadi-Trabelsi, M., Kalthoum-cherif, J., Vanderesse, R., Frochot, C., Guiavarc'h, Y., (2015). Accelerated solvent extraction of carotenoids from: Tunisian Kaki (*Diospyros kaki* L.), peach (*Prunus persica* L.) and apricot (*Prunus armeniaca* L.). *Food Chem.* 184, 131-139.

Zhang, Q.-A., Shen, Y., Fan, X.-H., García Martin, J.F., Wang, X., Song, Y., (2015). Free radical generation induced by ultrasound in red wine and model wine: An EPR spin-trapping study. *Ultrason. Sonochem.* 27, 96-101.

Zhang, Z.S., Wang, L.J., Li, D., Jiao, S.S., Chen, X.D., Mao, Z.H. (2008). Ultrasound-assisted extraction of oil from flaxseed. *Sep. Purif. Technol.*, 62, 192-198.

In: Applications of Ultrasound in the Beverage... ISBN: 978-1-63485-069-8
Editor: Juan Francisco García Martín

Chapter 4

NON THERMAL STERILIZATION APPROACHES BASED ON ULTRASOUND

Parag R. Gogate* and Paramjeet Khandpur
Chemical Engineering Department,
Institute of Chemical Technology,
Matunga, Mumbai, India

ABSTRACT

Vegetable and fruit juices have received considerable attention due to their content of health promoting compounds. Although thermal processing is generally used for controlling the safety standards, this approach gives disadvantages in terms of reduced quality of the processed juice as well as considerable processing costs. In recent years, non thermal approaches for sterilization have been developed with significant benefits and ultrasound is one of the emerging novel non thermal technologies. The current chapter will cover different aspects related to the application of ultrasound for sterilization of fruit and vegetable juices. The mechanism of microbial disinfection, available reactor designs and guidelines for important operating parameters will be discussed. Also comparison of the efficacy of ultrasound based approach with the conventional approaches will be discussed based on detailed analysis of the literature. The combined approach of ultrasound with ultraviolet irradiations and orange peel extract which can give synergistic results will

* Corresponding Author Email: pr.gogate@ictmumbai.edu.in.

also be highlighted. It has been established that the ultrasound based approach effectively maintains the quality, sensory and nutritional characteristics while preventing spoilage of juices.

Keywords: non thermal processing, reactor designs, thermal treatment, ultrasound, ultraviolet, sensory attributes nutrient quality, combined processing, disinfection, microbial growth, enzymes

INTRODUCTION

Food spoilage is a major cause of economic loss in the food industry and can be attributed to the growth of microorganisms or the activity of the enzymes present in the food leading to unwanted flavors or spoilage. Fruits and vegetables contain many essential nutrients, vitamins and minerals which are essential for normal growth and overall well-being of human beings. Considerable attention has been recently focused on vegetable and fruit juices due to the presence of health promoting compounds in the juices and ease of consumption especially for packaged juices. Juices play an important role in the daily nutrition and provide essential supplements. Juices are integral part of a healthy diet and are highly recommended for the nutritional content, phytochemical value and the presence of health promoting compounds. Juices are also rich sources of vitamin C, carotenoids, flavonoids, dietary fiber, phytochemicals etc. Ascorbic acid and water soluble polyphenols are the main antioxidants present in variety of juices [1–3] providing health promoting properties to the juices. These also contribute to the flavor and color attributes in the fruits and vegetables [4, 5] and any oxidation of polyphenols or the ascorbic acid can lead to browning of the juices. Microorganisms are the main agents responsible for food spoilage as well as for possible food poisoning. A number of food safety and food spoilage concerns are associated with fruit and vegetable juices. The outbreaks of food borne illness related to these products have increased the concerns of the public safety authorities.

Optimizing the use of fruits and vegetables in different product formats such as processed juice provides more options to the consumers. The use of processed food products also enhance the economic value of the produce since there is no problem related to perishable nature of fresh fruits and vegetables. Different treatment methodologies are required to extend the shelf life of juices. Heat treatment is the most widely used method for preservation of fruit and vegetable juices due to its effectiveness for microbial inactivation [6] although it has certain negative effects on the product nutritional and organoleptic value and may give cooked flavours [7, 8]. These problems with the thermal processing have led to the emergence of the so called "Nonthermal Technologies" [9] that can be effective at ambient or sub-lethal temperatures, thereby minimizing negative thermal effects on nutritional and quality parameters of food [10]. Non thermal processing approaches include the application of gamma irradiation, beta irradiation (electron beam), power ultrasound, ozonation, pulsed light, UV irradiation, pulsed electric field (PEF), high hydrostatic pressure etc. A lot of effort has been directed in developing new non-thermal techniques for food processing based on the use of lower thermal intensity and possibly ambient conditions. The technology, at the same time, must give significantly reduced microbial activity, ensuring safe consumption of food as well as higher shelf life for the product [11]. An optimum approach might have to be developed based on the balance of the microbial safety and maintaining the level of the food nutrients and fresh like attributes.

The quality of the processed food products is usually dictated by different standards such as U.S. Food and Drug Administration (FDA). The recommended guidelines for the processed juices are the 5 log reduction in the microbial content as per the FDA whereas different guidelines are available in terms of contents of antioxidants, vitamins, minerals, emulsifiers etc. There is need for concerted effort for accelerated development of non-thermal methods for quality maintenance of various fruit and vegetable products so that the health benefits of these products can be completely harnessed. Use of ultrasonic irradiations (either in direct contact mode or indirect mode) is a non-thermal approach for food preservation that can efficiently inactivate microorganisms [12] at less severe conditions as compared to the thermal processing which can possibly reduce the detrimental effects on nutrition, quality and sensory aspects of food. The present chapter will focus on providing an overview of application of ultrasound as a non-thermal preservation approach.

Ultrasound alone and/or in combination with other techniques has been reported to be effective against *Escherichia coli* in model fluids [13] as well as apple cider [11] and against *Listeria monocytogenes* in apple cider [14]. Ultrasound has also been reported to have lower effects on the quality of fruit juices such as orange juice [15, 16] guava juice [17] and strawberry juice [18]. Piyasena et al. [12] and Jiranek et al. [19] have reported an extensive analysis of the potential of ultrasound for inactivation of food borne pathogens. The combination of non-thermal methods is a new approach for preservation of fruit juices that can enhance the shelf life also giving faster microbiological reduction. Different combinations of ultrasound with other approaches such as ultraviolet irradiations, high pressure, chlorine dioxide, dense phase CO_2 and pulse electric field has been investigated in the recent past for inactivation of microbes and enzymes [20–24]. It is important to understand that combination of non-thermal processing like ultrasound and ultraviolet irradiations might lead to some negative effects on the nutrient quality in comparison with ultrasound alone but will definitely lead to enhanced deactivation and hence enhanced shelf life. An optimized process and detailed analysis should be able to help in maintaining the proper balance and the present chapter can be a useful guideline for this optimization.

MECHANISM OF DISINFECTION AND FUNDAMENTALS OF COMBINATION APPROACHES

Use of ultrasound (mostly over frequency range of 20–200 kHz) offers good promise for many applications in food industries and it has also shown potential in food sterilization yielding the desired microbial safety levels. The microbial destruction effect of ultrasound has been attributed to the cavitation phenomenon. Cavitation refers to the formation of bubbles or cavities in liquids, subsequent growth under the oscillating pressure fluctuations and finally violent collapse of the bubbles producing intense localized changes in pressure and temperature, coupled with significant shear and turbulence. The shear-induced breakdown of cell walls and DNA damage via free radical production from thermolytic dissociation leads to overall deactivation [25]. Inactivation of pathogenic and spoilage microorganisms or enzymes by sonication is mainly caused by the physical effects (turbulence, liquid circulation, shear or micromechanical shocks) and/or chemical effects (formation of free radicals such as H• and OH• radicals by the decomposition of water inside the cavities) of cavitation [12, 26-27]. The exact controlling mechanism will be dependent on the operating conditions, more importantly the operating frequency and the power dissipation levels. It is important to note that controlled application of ultrasound is required so as to maximize the degree of inactivation with as low as possible reduction in the nutrient quality. A similar analogy can be given in terms of application for intensification of enzymatic reactions where controlled application is required as stimulus to the enzyme rather than deactivating the enzyme itself. Thus there is also need to optimize the different parameters such as power dissipation, treatment time and duty cycle that critically decide the cavitational intensity in the processing.

Combination of ultrasound with other preservation processes such as heat, high pressure, or ultraviolet irradiations has also been shown to have good potential for industrial applications [6]. Ultraviolet light treatment is a relatively new process that is becoming more common in the food processing industry, as it holds considerable promise as an alternative to traditional thermal processing [28]. Along with the ultrasound, use of ultraviolet irradiations is also gaining momentum in food processing for its beneficial anti-microbial properties and possible lower effect on the nutrients as compared to thermal sterilization [29, 30]. UV light has already been used in food industries for fruit juice disinfection [31] and vegetable surface decontamination applications. FDA has also approved UV irradiation as a safe food sanitation technology that could be used for food and beverage sterilization applications. The UV germicidal effect is based on the absorption of light photons by exposed cells and damage to the DNA due to thymine dimer formation, which destroys the ability of microorganisms to grow [32, 33]. The wavelength of 254 nm (UV-C) has the highest germicidal effect and is used for disinfection of surfaces, water and certain food products. The inactivation mechanism depends on the absorption of UV photons by the genetic materials and subsequently the formation of dimers which inhibit the transcription and replication of the cell [30, 34]. It is also important to note here that UV light offers some problems in the processing based on lower penetration levels and inefficient distribution especially for juices containing solids and at larger scale operation. Several factors such as the type and power of lamp used, density, viscosity, turbidity, absorbance coefficient of food material being treated will help determine the efficiency of UV disinfection. Combining ultrasound in a hurdle technology with ultraviolet treatment can potentially enhance the overall quality of minimally processed foods as this can reduce the treatment times and can work in complimentary manner with the UV irradiations. Some researchers have investigated the effects of ultrasound on juice shelf life and optimized the extent of disinfection as a function of different operating parameters [15, 35-37]. There has been significant work related to understanding the effect of operating parameters on the microbial log reduction using UV or ultrasonic irradiations [38–43].

FUNDAMENTALS OF REACTOR DESIGNS

An ultrasonic system consists of two essential components: a medium through which the sound travels to generate cavitation and a transducer which is the source of high energy vibrations. In general, ultrasonic transducers are designed to convert either mechanical or electrical energy into high frequency sound. Electrical energy at low frequency (50-60 Hz) is first transformed to a high frequency (i.e., 20 kHz) corresponding to the frequency of the ultrasound by a generator, and subsequently to mechanical vibration via the piezoelectric crystals in the transducer, which is then amplified by a probe to obtain certain level of ultrasound intensity. The generator provides electrical energy from main voltage to power an ultrasonic device with the set frequency which is transferred to the transducer. The transducer converts electrical oscillations into mechanical vibration and is an interface between the generator and the sonotrode [44]. The sonotrode carries the vibrations from the transducer to the suspension to produce intense cavitation [45]. The ultrasound device generates heat which is dissipated into the suspension and sometimes the heat can affect the quality and the accuracy of the experiments and needs to be controlled. A means of a temperature control such as an ice or water bath system can be included so that there would be no thermal effect and the suspension temperature is maintained sufficiently to help in maintaining the quality of food products.

The most common design for the ultrasound based reactors is the ultrasonic horn that is most commonly used for laboratory scale characterization studies. These are typically single transducer based designs delivering significant magnitude of power directly to the processing area. The liquid circulation generated due to the dissipation of energy can be powerful to provide efficient mixing at smaller scale of operation and also the levels of turbulence generated can be sufficient for particle breakage. Another commonly used design is the ultrasonic bath, which has mostly multiple transducers attached to the bottom of the cleaning tank. The active zone of cavitation is restricted to a vertical plane above the transducers and hence the area of irradiating surface as well as the arrangement of transducers needs to be adjusted so as to get better distribution of the incident energy as well as the cavitational activity distribution. The reaction vessel can be immersed in the coupling fluid contained in the bath (indirect sonication) or the bath itself can be used as the reaction vessel but may require additional mechanical agitation for uniform mixing, especially for large scale operations.

For large scale operations typically continuous processing is required and configurations based on the flow systems involving multiple transducers are needed. A simple flow system can be constructed based on the use of many ultrasonic probes in the flow loop and the number of transducers can be decided based on the treatment times required for the specific application. Utilizing probe in the flow loop also gives the flexibility of using high intensity ultrasound which may be necessary if the microbial loading is on higher side or the system to be treated is complex, i.e., involving large solid loadings. Since probe would be in direct contact with the liquid to be treated, this type of arrangement would also suffer from general disadvantages of the probe such as tip erosion and corrosion that can give product contamination.

Recent developments into transducers and the bonding mediums have allowed manufacturing of the reactor vessels with many transducers attached to the reactor walls or to the bottom of the reactor. Such an arrangement also gives the flexibility of adjusting the geometry of the reactor and number of transducers so as to control the cavitational intensity and the activity distribution in the reactor.

Reactors based on the combined use of ultrasound and ultraviolet irradiations [46] can also be developed as this can yield synergistic effects for microbial deactivation also possibly giving intensified processing. A simple system can be based on the use of ultrasonic horn and an additional effect can be created by exposing the juice sample to ultraviolet light simultaneously along with ultrasound (Figure 1). It is important to use a quartz beaker as any other material can lead to significant attenuation of the incident ultraviolet irradiations. During the processing it might be recommended to use filtered juices as the presence of solid particles are expected to provide a protective shield for the microorganisms as well as reduce the available intensity of the incident light for treatment [47]. An alternative option would be to use enhanced intensities of ultrasound so that the protective shields are broken based on the turbulence effects. However, this can also have a negative effect on the nutrient quality.

A simple pilot scale reactor used in our earlier work [46] has now been described for better understanding on the reactor designs and the working methodology for investigating the efficacy for microbial deactivation and assessing the effects on nutrient quality. The ultrasonic reactor used in the work has a working capacity of 7 L, power dissipation of 150W and operates at a fixed frequency of 36 kHz. Generally speaking, ultrasonic reactors can be operated at fixed frequency and variable power dissipations. The internal body of the reactor is made of stainless steel, which gives better processing effects especially for juices with acidic pH. It is also imperative to understand that direct contact of the ultrasonic horn with the processed juice can lead to possible contamination with metal particles and hence it is an important requirement to adjust the processing conditions. The material of construction of both the reactor and horn used for this purpose can be customized when used in food industries and special food grade materials and alloys can be used. Mawson et al. [48] have also reported that under optimized processing conditions, metallic nanoparticles were not detected within the size range of concern to health, therefore suggesting the safety of using sonotrodes and transducers in direct contact with food materials. The reactor used in our work [46] was equipped with a single transducer placed horizontally at the bottom of the reactor. The reactor was also provided with facility to monitor the temperature of the juice during the processing. Preliminary studies performed in the laboratory established that the efficacy of ultrasound was reduced significantly when the juice samples were not filtered as the presence of solids had a protective effect on the microorganisms. To get similar degree of microbial log reduction, the required sonication time needed to be increased, though it also meant a negative effect on the nutrients present in the juices as established by the nutrient quality analysis.

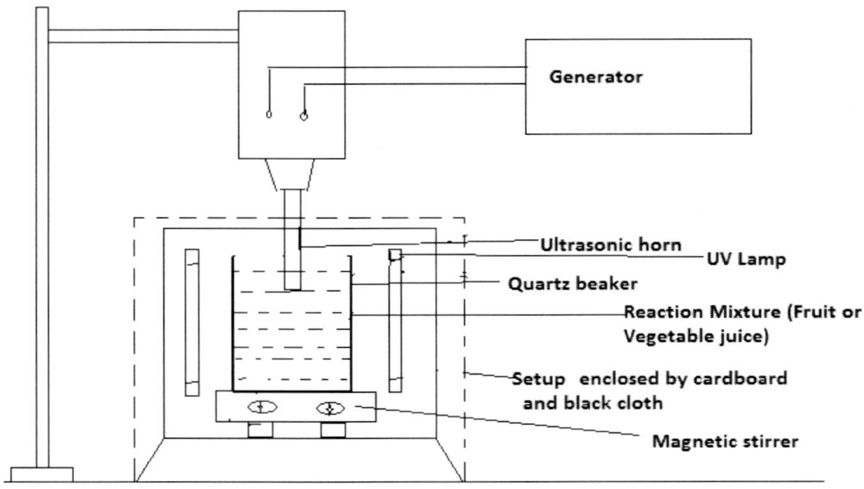

Figure 1. Ultrasonic - UV reactor consisting of a probe system consisting of a probe (or sonic horn), a transducer, generator and UV lamps.

Guidelines for Operating Parameters

The effectiveness of a power ultrasound treatment is influenced by a number of factors, including the frequency and intensity of ultrasound, amplitude of the ultrasonic waves, the exposure/contact time, the type of microorganism, cell conditions (non-adapted, acid adapted), fruit juice type, presence of pulp, viscosity, volume of food to be processed, the composition of the food, solvent, gas type and content in working medium, treatment temperature, geometry of the reactor, uniformity of the acoustic field in the treatment chamber, and externally applied pressure. Optimization of operating parameters for ultrasound processing of juice is mandatory so that maximum log reduction is obtained at the same time maintaining the juice quality. We now present some guidelines for the important operating parameters.

Frequency of Ultrasound Waves

In general, increasing ultrasonic frequency results in a decrease in the intensity of cavitation in liquids considering the physical effects of turbulence and shear. The formation of cavitation bubbles within a liquid is more difficult at higher frequency. Therefore, as the frequency increases, bacterial killing decreases. The use of low frequency irradiation is generally recommended due to the dominant physical effects and its controlling action on the deactivation.

Amplitude of Ultrasound Waves (USW)

Amplitude of USW has also been found to have a major effect on inactivation potential of ultrasound as the amplitude of the ultrasound determines the intensity of cavitation. The higher the amplitude of the sound waves, higher will be the intensity of cavitation and hence better deactivation efficiency can be obtained. However the negative effect on the stability of transducers should be established before using maximum amplitude.

Energy of Ultrasound Waves

The energy output of ultrasound can be given in terms of intensity (Wcm^{-2}) or acoustic energy density (Wml^{-1}). The inactivation of bacteria has generally been found to increase with an increase in intensity of ultrasound. Intensity of sonication is proportional to the amplitude of ultrasonic vibration, which results in an increase in the sonochemical effect with increasing amplitude [49]. Again there would be optimum power dissipation for maximum effects which need to be established for the specific system under question.

Temperature

Temperature is an important parameter for sonication as it has a great influence on the intensity of cavitation. It also modifies several properties of the liquid medium, such as viscosity, surface tension and mainly vapor pressure, which influence the onset of cavitation as well as the cavitational intensity [50]. Although an increase in temperature leads to an increase in the number of cavitation bubbles, intensity of the collapse of these bubbles is hindered by the increased vapor pressure. Thus cavitation is better attained with higher intensity at lower temperatures. Contrarily, the temperature also has an effect on viscosity where an increase in temperature reduces the viscosity allowing more violent cavitation bubbles [50, 51]. With specific context to the food processing, using higher temperature would mean that better deactivation is obtained due to the thermal effects though it would also have a negative effect on the nutrient quality as well as the sensory attribute.

Operating Pressure

The external pressure is also a critical factor affecting the intensity of cavitation. An increase in pressure leads to an increase in both the cavitation threshold and the intensity of bubble collapse. At high pressure, the time required for bubble collapse decreases and the intensity of implosion increases leading to more inactivation at higher pressures [27]. Using higher pressures would also result in higher deactivation efficiency and lead to intensified processing.

Composition of Foods

The characteristics of the food or substrate can influence the effectiveness of the ultrasound treatment. Differences in effectiveness may be due to intrinsic effects of the medium on the ultrasound action (cavitation) or due to changes in ultrasound penetration and energy distribution [52]. Owing to their heterogeneous nature based on the presence of solid particles, foods exert a protective effect on microbes when they are subjected to inactivation treatments such as heat, chemical and high pressure treatments [53, 54] and similar effect would be expected in the case of ultrasound. Thus, some pretreatment for removal of solids may be required. On the contrary, the negative effect during the treatment can also be compensated by using higher intensity processing.

Viscosity

Viscosity of the liquid medium also affects cavitation. In highly viscous medium, ultrasound diffusion is impeded thus reducing the degree to which cavitation occurs. In this case, lower frequency and high power ultrasound is more effective in penetrating the viscous product as compared to the higher frequency ultrasound, which is easily dispersed within the medium.

Presence of Gas

Monoatomic gases like He, Ar and Ne, when dissolved in a liquid, favor the sonication process by acting as nuclei for cavitation [45]. Thus, in order to increase the cavitation effect, the gas must be bubbled continuously into the solvent.

Water Activity

Water activity of the medium affects the inactivation of microorganisms; however sonication shows lesser impact as compared to the heat treatment which has been established based on a specific case study observed in the literature. It has been reported that the D60°C value for heat inactivation of *Salmonella Enteritidis* increased from 0.10 to 2.70 min (27 times) when the water activity decreased from >0.99 to 0.96. However, the degree of disinfection for sonication increased from 0.89 to 1.37 min (~1.5 times) for a same decrease in water activity. The observed results led to a significant conclusion that while a wide range of heat treatments would be required for processing of foods with different water activities, one sonication treatment would be sufficient to inactivate bacteria in a wide range of food products [55].

Operating pH

Inactivation of microorganisms by ultrasound waves is also influenced by the pH of the medium and in general acidic conditions are more suitable. Mack and Roberts [56] reported that as the pH of malic and citric acid solution was decreased from 4.0 to 2.5, the log reduction of *E. coli* using sonication increased from 0.6 to 1.0 log.

Presence of Chemicals and Antimicrobials

Combined use of chemical and antimicrobial agents with ultrasound treatment has been reported to give significant increase in the microbial deactivation efficiency. Ferrante et al. [57] reported that 20 kHz ultrasound alone was able to reduce the number of *L. monocytogenes* in orange juice by ~1 log in 15 min. However, sonication of *L. monocytogenes* in orange juice supplemented with vanillin (a natural antimicrobial substance) at concentrations of 1000 and 1500 ppm, resulted in 1.8 and 3.5 log reductions, respectively, in similar treatment times.

Type of Microorganisms

The inactivation mechanism of ultrasound is mainly due to direct damage to microbial cell membranes [58]. Bacterial cells differ in their sensitivity to ultrasound treatment [59] based on size, shape and species [60]. Drakopoulou et al. [61] reported that in the disinfection of wastewater using sonoporation, among vegetative forms, gram positive cells were more resistant to ultrasound as compared to the gram negative cells. Also aerobic bacteria show more sensitivity as compared to the anaerobic bacteria. Generally, bacterial spores are the most resistant microorganisms to any physical or chemical treatment and hence extended periods of sonication would be required to render a product safe. Also cocci (spherical shaped cells) are more resistant than rod forms [62, 63] and bigger cells show more inactivation than smaller ones.

Exposure Time and Cell Density

Typically the fraction of the undamaged cells decreases with an increase in the exposure time to ultrasound. The effectiveness of the ultrasound in the destruction of the cells also decreases with an increase in the concentration of the cells in the sample [64]. It has been also reported that these effects were more prominent for large volumes of the sample.

OVERVIEW OF LITERATURE FOR DEACTIVATION USING ULTRASOUND AND ULTRAVIOLET IRRADIATIONS

Development of commercially feasible method(s) to minimize the microbial contamination, nutritional component loss, and occurrence of offensive flavor and odor components in beverages is very important. The use of ultrasound within the food industry has been a subject of research and development for many years especially for process control and characterization of physicochemical properties of food products. The use of ultrasound waves to inactivate microorganisms is comparatively recent. Application of ultrasound is a non-thermal technology which contributes to the enhanced microbial safety also giving higher shelf life, especially in foods with heat-sensitive, nutritional, sensory, and functional characteristics [65]. Similar to the conventional heat treatment, ultrasound can inactivate a wider range of bacterial species. Ultrasound can also be combined with heat (thermosonication) and pressure (manosonication) for the inactivation of microorganisms as well as heat resistant enzymes at higher rates [66, 67]. The primary antimicrobial effects and the driving force of ultrasonication are attributed to acoustic cavitation effects leading to an increase in the permeability of membranes due to the thinning of cell membranes [68], localized heating [27], and production of free radicals [69, 70]. Ultrasound has been commonly applied for the inactivation of different pathogenic and spoilage micro-organisms in a variety of juices such as apple cider [71], orange [72-74], guava [75] and tomato [76]. A comprehensive overview of the different applications for juice preservation has been presented in Table 1 whereas the important discussion related to key aspects and guidelines have been presented now.

Ultrasound seems to be promising in extending the shelf life of juices as reported in the study related to the reduction in the activity of pectinmethylesterase (PME) in lemon juice thereby improving cloud stability when stored for 18 days at 4ºC [66]. Ultrasound has also been reported to improve the functional properties of foods as well as in retaining nutritional and organoleptic properties of juices. Ugarte et al. [11] demonstrated that the physical properties of apple cider such as titratable acidity, pH and °Brix were not affected by ultrasound treatment. Also degradation of ascorbic acid in orange juice during storage at 20ºC has been reported to be lesser in the case of thermosonication as compared to the heat treatment alone [66]. Ultrasound has been shown to be effective against various enzymes including thermostable enzymes. The enzyme activity of lysozyme was found to decrease by ~16-fold (from 100% residual activity to 6% residual activity) in 30 sec upon ultrasound assisted heat and pressure treatment at 80°C and 200 kPa in phosphate buffer. Use of ultrasound also allows maintaining the quality parameters as compared to the thermal processing. Dubrovic et al. [21] investigated the effect of high intensity ultrasound on the anthocyanin content in strawberry juice and compared the results with the conventional thermal processing. It has been reported that ultrasound treatment can be a good alternative to the pasteurization as the total anthocyanin content is preserved to a higher extent in the ultrasound processing as compared to the thermal pasteurization.

Various combinations of ultrasound with heat and pressure can also be applied with beneficial results; however the operating conditions need to be optimized. In thermosonication, the product is subjected to ultrasound and moderate heat simultaneously and the combination approach shows the same inactivation level compared to the treatment without ultrasound at high temperature. Using too high temperature in combination with ultrasound would also lead to negative effects on the cavitational intensity and needs to be avoided. The low effectiveness at higher temperatures is attributed to the decrease in the intensity of bubble collapse due to the higher vapor pressure which results in cushioning effect during the collapse. It has also been reported that bacterial cells generally become more sensitive to heat treatment after being subjected to ultrasound treatment. Manosonication provides the possibility to inactivate enzymes and/or microorganisms by combining ultrasound with moderate pressure pulse at 100 - 300 kPa and giving advantage of use of low temperature. Increased lethality under moderate pressure was suggested to be the effect of increased intensity of cavitation. An increased pressure beyond the moderate level does not provide an increased rate of inactivation. It is also important to note that the pressure level employed in manosonication is not in the lethal range as that used in high pressure processing.

The short-wave ultraviolet (UV-C) light is also a physical method that can be applied for water treatment and surface disinfection due to the lethal effects on microorganisms such as bacteria, viruses, protozoa, yeasts, and molds [77-79]. The amount of cell damage depends on the dose of UV energy absorbed by the microorganisms and their resistance to UV. The U.S. FDA has approved the use of UV-C light in the processing of fruit juices to reduce the microbial load. UV-C also gives the advantage of no formation of toxic by-products during the treatment and requirement of lesser energy as compared to thermal pasteurization processes, and at the same time maintaining maximum aroma and color of the treated fruits [80]. A brief overview of the UV-C treatment for reducing the microbial load in different fruit juices and nectars has been presented in Table 2 while some important results have been presented here so as to understand the basis for possible combination with ultrasound. Under suitable treatment conditions, more than 5-log reduction of some pathogenic microorganisms, such as *E. coli*, in fruit juices has been reported by the use of UV irradiations [29, 33]. The required UV dosage is strongly dependent on the fruit juice characteristics. It was reported that the minimum treatment condition for clear apple juice was UV dosage of 230 J L^{-1}, whereas higher UV dosage levels was needed for cloudy juices such as orange juice and tropical juices [29].

Choi and Nielsen [81] demonstrated that UV pasteurized apple cider was superior in color and overall sensory scores compared to thermally pasteurized apple cider. UV-irradiated samples were lower in soluble solids for the first 7 days and showed no significant difference in consumer acceptability [81]. Studies using apple cider that utilized the CiderSure 3500 UV apparatus (FPE, Inc., Rochester, NY) confirmed the ability of the apparatus to achieve a 5-log reduction of *Cryptosporidium parvuum* and *E. coli* [82, 83]. The apparatus was designed to allow the apple cider to pass through a series of eight germicidal UV lamps in thin films. The wavelength and intensity inside the CiderSure apparatus was 254 nm and 14.3 mJ/cm^2 of UV irradiation, respectively, with exposure times between 1.2 and 1.9 sec [82].

Table 1. Examples of ultrasound induced microbial inactivation of fruit and vegetable juices

Beverage	Microorganism	Treatment Parameters	Log Reduction	References
Carrot juice	*E.coli*	19.3 kHz, 700-800 W, 1 min, 60°C	2.5	Zenker et al. [90]
Apple cider	*L. monocytogenes*	48 kHz, 600 W, 3 min, 25°C	1 to 2	Rodgers and Rysers [91]
Apple cider	*E. coli* O157:H7	48 kHz, 600 W, 5 min, 25°C	1 to 2	Rodgers and Rysers [91]
Apple cider	*L. monocytogenes*	20 kHz, 750 W, 5 min, 0.46 W/mL, 60°C	5	Baumann et al. [14]
Apple juice	*S. cerevisiae* and *L. monocytogenes*	20 kHz, 5-10 min, 95 μm amplitude, 35°C in combination with UV lamp (30 W, 254 nm, low pressure)	5	López- Malo et al. [92]
Apple cider	*E.coli* O157:H7	20 kHz, 150 W, 18 min, 118 W/cm^2, 57°C	6	D'Amico et al. [93]
Apple cider	*E. coli O157:H7*	20 kHz, 0.46 W/ml for 45°C, 50°C, 55°C, and 60°C	5.3 >5	Ugarte-Romero et al. [11]
Orange juice	Total mesophilic aerobes	500kHz, 240W, 15 min, 60°C	3.4	Valero et al. [15]
Guava juice	Aerobic plate Count	Sonication and carbonation 35 kHz (sonication bath); 30 min	<1	Cheng et al. [75]
Orange juice	*L. monocytogenes*	45°C, high-intensity ultrasound (600 W, 20 kHz, 95.2-μm wave amplitude), and the addition of different levels of vanillin (0, 1,000, 1,500 and 2,000 ppm), citral (0, 75, and 100 ppm)	TDT: 1.6 and 2.6 min	Ferrante et al. [57]
Model juice (simulated berry juice and apple juice)	*E. Coli* ATCC 25922	24 kHz, 100 μm amplitude, 3-4 min, 85 W/cm^2	5	Salleh-Mack and Roberts [13]
Orange juice	*S. cerevisiae*	20 kHz, amplitude at ≥40% (≥80 W), 50°C, 150.38/cm^2, 15 min	6	Kuldiloke et al. [94]

Beverage	***Microorganism***	**Treatment Parameters**	**Log Reduction**	**References**
Apple juice	*Alicyclobacillus acidoterrestris*	24kHz, 300W, 60 min, duty cycle of 5 s on and 3 s off	80%	Yuan et al. [95]
Model orange juice and model apple juice	*E. Coli* 25922 *E. Coli* NCTC 12900	20 kHz, 37.5 μm amplitude with pulse durations of 5 s on and 5 s off, 11.1 min, T > 30ºC 20 kHz, 37.5 μm amplitude with pulse durations of 5 s on and 5 s off,13.8 min, T > 30ºC	5	Patil et al. [73]
Orange juice	Aerobic mesophilic count (AMC)	20 kHz, 500 W, 8 min, 89.25μm amplitude,10ºC	1.38	Gomez Lopez [74]
Orange juice	Yeast and Mold count(YMC)	20 kHz, 500 W, 8 min, 89.25 μm amplitude, 10ºC	0.56	Gomez Lopez [74]
Orange juice	*E. coli* O157:H7 *L. monocytogenes* *E. coli* O157:H7 *L. monocytogenes*	20 kHz, 750 W, 99.2 μm amplitude, 5 min, 45ºC 20 kHz, 750 W, 99.2 μm amplitude, 5 min, 50ºC 20 kHz, 750 W, 99.2 μm amplitude, t < 30 s, 45ºC in combination with eugenol and linalool (1000 μl/l) 20 kHz, 750 W, 99.2 μm amplitude, t < 30 s, 50ºC in combination with essential oils such as eugenol and linalool (1 ml/l)	1.3 1.3 5.92 and 5.85, respectively 5.53 and 5.71, respectively	Utkun et al. [96]
Kiwi juice	Total bacteria	Sonication and carbonation 35 kHz (probe sonication); 30 min	5	Dizadji et al. [97]
Apple-carrot blend juice	Total aerobic viable count	20 kHz, 1000 W, 10 min, 25°C ± 1ºC.	3 to 3.5	Gao et al. [98]
Cactus pear juice	Total plate count and Enterobacteria	20 kHz, 1500 W with 40%, 60% and 80% amplitude levels and time (10, 15 and 25 min), pulse durations of 2 s on and 4 s off.	2	Zafra-Rojas et al. [99]
Pomegranate juice	*E. Coli* and *S. cerevisiae*	20 kHz, 24.4–61 μm amplitude level (100%), 25 ± 1ºC, and 15 min	3.47 and 1.86, respectively	Alighourchi et al. [100]

Table 2. Examples showing effect of ultraviolet light on quality attributes of fruit products

Beverage	Treatment Parameters	Quality Changes	Sources
Apple cider	14 mJ/cm^2 and 254 nm	Significant changes in the pH, titrable acidity, soluble solids and turbidity of UV-irradiated samples during storage. An acceptable reduction in microbial loads.	Tandon et al. [101]
Mango nectar	450 kJ/m^2, 254 nm, 30 min	Yellow and yellow-orange color of mango nectar was maintained after 26 days of storage, and its shelf life was extended to 20 days with no microbe growth.	Guerrero-Beltran and Barbosa-Canovas [31]
Apple cider	254.7 nm, 35.1 mJ/cm^2, 8.12 s	UV treated apple cider stored at 4 ^{0}C had a shelf life 7 days longer than untreated apple cider.	Donahue et al. [34]
Orange juice	73.8 mJ/cm^2 at 254 nm wavelength	Exposed orange juice to a limited UV dose and succeeded in extending its shelf life from 2 days to more than 5 days	Tran and Farid [80]
Apple cider	254nm, 10min	UV pasteurized apple cider was superior in color and overall sensory scores compared to thermally pasteurized apple cider. UV-irradiated samples were lower in soluble solids for the first 7 days and showed no significant difference in consumer acceptability	Choi and Nielsen [81]

The efficacy of the disinfection system depends on the absorbance of the medium, moisture content, amount of solid particles and suspended materials, flow rate of the fluid, reactor design, UV intensity which is related to the age and rating of lamps used, exposure time, type of microorganisms, growth phase of the organism, and initial microbial density [79]. Caron et al. [84] suggested that suspended particles and dissolved matters increase the absorbance of the medium. Increased absorbance of the liquid being irradiated reduces the effectiveness of the process [85]. Also, particulate matters can reduce the efficiency of UV by scattering of the incident light [84]. In addition to the attenuation of UV light, suspended particles provide a protective shield for the microorganisms [30]. It was also reported that the order of resistance of microorganisms against UV-C light follows an increasing order starting with bacteria, viruses, fungi, spores and cysts [86]. Since yeasts and molds contain less pyrimidine bases, they are indicated to be more resistant to the UV-C light disinfection process. Koutchma et al. [87] also highlighted the importance of the proper reactor design to obtain the required inactivation efficiency of the system. The use of turbulent flow and extremely thin film was suggested to enhance the effectiveness of UV-C light in treating the liquid foods. The physical, chemical and optical properties of food material being treated (density, viscosity, turbidity, absorbance coefficient) [88] also affect the inactivation efficiency. Due to all these issues, combination with other treatment approaches is a more viable option as compared to the use of ultraviolet irradiation alone.

Combined Ultrasound and Ultraviolet Irradiations

The combination of ultrasound with some non-thermal and/or physical–biological methods is an attractive approach to enhance microbial inactivation [22]. Combination of ultrasound with other preservation process such as heat, high pressure, or ultraviolet irradiations has been shown to have good potential for industrial applications [28]. Combining ultrasound in a hurdle technology with ultraviolet treatment can potentially enhance the overall quality of minimally processed foods as this can reduce the treatment time and can work in complimentary manner with the UV irradiations. Combined treatment of ultrasound and ultraviolet has better microbial inactivation rate in juices as compared to ultrasound alone. Though a lot of work has been done on the application of ultrasound or ultraviolet irradiations during the processing of food, the combined effects of ultrasound and UV on fruit and vegetable juices have been depicted in only few literature reports. Char et al. [20] investigated the application of combined processing of orange and apple juice involving high intensity ultrasound and UV-C light and reported that the combined treatment was more effective for inactivation of the *E. coli*, *Saccharomyces cerevisiae*, and a yeast cocktail. It is important to understand that combination of non-thermal processing like ultrasound and ultraviolet irradiations might lead to some negative effects on the nutrient quality in comparison with ultrasound alone but will definitely lead to enhanced deactivation and hence enhanced shelf life. An optimized process and detailed analysis should be able to help in maintaining the proper balance. One such study was recently reported by our group on different fruit and vegetable juices. Khandpur et al.

[46] investigated the effect of novel approaches based on the use of ultrasound and ultraviolet irradiations on the nutritional quality of different fruit and vegetable juices (orange, sweet lime, carrot and spinach juices) and its comparison with the conventional thermal pasteurization operated at 80°C for 10 min. The ultrasound sterilization parameters were maintained as ultrasound frequency of 20 kHz and power of 100 W for 15 min. For the case of ultraviolet irradiations, 2 UVC lamps (254nm) of 8W were placed in parallel on either sides of the reactor. The treated juices were analyzed for total phenol content, antioxidant activity, vitamin C, carbohydrates etc. It has been established that ultrasound processed juice retained most of the nutrient components to higher extent in comparison to all the other techniques used in the work. Combination of ultrasound and ultraviolet irradiations used to achieve an effective decontamination of juices (recommended 5 log reduction of microorganisms) also retained nutrients to a higher level in comparison to the thermal method; however some losses were observed as compared to the use of only ultrasound which could be attributed to inefficient heat exchange. A scale up attempt was also made for treatment of spinach juice using ultrasonic reactors and analysis for quality attributes confirmed that the juice satisfied the requirement of desired nutrient contents for 18 days shelf life trial in refrigerated storage conditions. The work has clearly established the usefulness of ultrasound based treatment in maintaining the nutritional quality of beverages while giving enhanced shelf life as compared to the conventional approaches.

Ultrasound and Orange Peel Essential Oils

An antimicrobial is a substance which kills or inhibits the growth of many types of microorganisms such as bacteria, fungi or protozoan's. Citrus fruit products act as antimicrobial agents against the bacteria and the fungus. The waste products obtained during the processing of citrus fruits for juice production can be a sustainable source for recovery of many important natural active ingredients. For example, significant quantum of flavanones and many polymethoxylated flavones are present in the peel of citrus fruits and these flavonoids offer significant prospective applications mainly due to the range of biological activities. Citrus peel is a natural antimicrobial and also rich in polyphenol compounds, mainly flavonoids. Flavonoids can function as direct antioxidants and free radical scavengers, and can modulate enzymatic activities [102]. Citrus flavonoids have a large spectrum of biological activity including antibacterial, antifungal, antidiabetic, anticancer and antiviral activities [103, 104]. These compounds have commercial interest because of their multitude applications in the food and pharmaceutical industries. They also contain a wide variety of secondary metabolites, such as tannins, terpenoids, alkaloids and flavonoids including hesperidin, naringin, nobiletin, anthocyanins, and coumarins located in the cell vacuoles of the citrus peel albedo tissue [105] that are found to have effective antimicrobial properties. The monoterpenes affect the structural and functional properties of lipid fraction of the plasma membranes of bacteria and yeasts, causing leakage of intercellular material and exit of critical molecules and ions leading to death of microbes. Citrus peel extracts are usually composed of citrus peel oils. Plant derived oils are the products of secondary metabolisms and have substantial amount of antimicrobial activity on many species. Citrus essential oils have been found to inhibit Gram +ve and Gram –ve bacteria as well as yeasts, molds and food poisoning bacteria [106].

A novel approach of using the extracted antimicrobial metabolites from the citrus peel wastes in combination with ultrasound for juice preservation was reported in our recent work [46]. The combined study of ultrasound and crude extract of orange peel essential oil was carried out for the specific case of orange juice. The study targeted the extraction of antimicrobial metabolites from the peels. The orange peels were dried and antimicrobial metabolites were extracted using the solvent extraction procedure. Extracts were subjected to antibacterial and antifungal susceptibility assay using the agar disc diffusion method. Later crude essential oils extracted from orange peels were combined with ultrasound treatment under optimized conditions and this hurdle treatment concept has been evaluated as a sterilization approach focusing on the effects on the microbial and nutrient parameters during the storage. Shelf life study was also performed by carrying out analysis at regular intervals for a maximum period of 8 weeks.

Studies related to only orange peel extract revealed that different solvent extracts of orange peel oil caused different inhibition zones on the tested bacteria. All the solvent extracts showed antibacterial activity to some extent, however *E.Coli* was more susceptible to methanol and hexane extracts which gave the highest zone of inhibition. Enterobacter was found to be sensitive to hexane extract. *Pseudomona Aeruginosa* was found to be the most sensitive against ethanol and chloroform extract showing a wider zone of inhibition. Growth of *Bacillus subtillis* was inhibited by acetone crude extract. Antifungal activity was also exhibited by these crude extracts. Hexane showed higher resistance to the growth of *Aspergillus Niger.* It was also established that better action of the oils was observed in the citrus juice samples as compared to the vegetable juices which can be attributed to the higher disinfection efficiency under acidic conditions. It was also established that more than 5 log reduction is achieved using the combination of sonication along with the peel oil extracts at faster rates. It was concluded that the formulation of essential oils can be added to fresh juices to reduce the degree of bacterial contamination and can help in preventing outbreaks of infective diseases as well as improve the general health conditions of people.

CONCLUSION

The aim of the current chapter was to demonstrate the feasibility of applying novel techniques, based on ultrasound induced cavitation along with hurdles such as natural antimicrobials and UV light, to eliminate the microbiological hazards associated with vegetable and fruit juices. The study also highlights important issues related to the effects of generally used thermal processing on the juice quality in comparison to the novel technique of ultrasound. The quality criteria were based on the most important nutrient and sensory attributes of fruit and vegetable juices. Based on the analysis for the microbial content presented in different studies overviewed in the chapter, it can established that there is no significant difference between the thermal and non-thermal treatments and the desired FDA log reduction could be obtained after optimization of the ultrasound parameters. Studies related to the nutrient analysis have demonstrated that the ultrasound treated juice showed much better results as compared to the thermal processing and the processed samples using ultrasound were at par with the fresh juice. Sensory characteristics of food are of utmost importance as it may contribute significantly to the consumer acceptance or rejection of the product. Sensory property analysis of the ultrasound treated juice presented in the chapter established that the samples were acceptable to the consumers and rated at par with the fresh juice. The reported results regarding consumer preference strongly suggest that ultrasound technology has a potential area of application in the food industry as the ultrasound treated juice might be the best option for consumers preferring health benefits from their food along with taste and fresh like attributes. Overall, it has been established that the non-thermal techniques based on the use of ultrasound could be helpful for food industry to produce fruit juice with high nutrient quality and paving the way for large scale commercialization of the technology.

REFERENCES

[1] M. Su, P. Chien, Antioxidant activity, anthocyanins, and phenolics of rabbiteye blueberries (vacciniumashei) fluid products as affected by fermentation, *Food Chem.* 104 (2007) 182–187.

[2] E. Nicoue, S. Savard, K. Belkacemi, Anthocyanins in wild blueberries of Quebec: extraction and identification, *J. Agric. Food Chem.* 55 (2007) 5626–5635.

[3] J. Piljac-Zegarac, L. Valek, S. Martinez, A. Belscak, Fluctuations in the phenolic content and anitioxidant capacity of dark fruit juices in refrigerated storage, *Food Chem.* 113 (2009) 394–400.

[4] C. Kaur, H.C. Kapoor, Antioxidants in fruits and vegetables – the millennium's health, *Int. J. Food Sci. Technol.* 36 (2001) 703–725.

[5] H.J. Heo, Y.J. Kim, D. Chung, D.O. Kim, Antioxidant capacities of individual and combined phenolics in a model system, *Food Chem.* 104 (2007) 87–92.

[6] A.A.L. Tribst, M.A. Franchi, M. Cristianini, Ultra-high pressure homogenization treatment combined with lysozyme for controlling *Lactobacillus brevis* contamination in model system. *Innovative Food Sci. Emerging Tech.* 9 (2008) 265–271.

[7] J.F. Vachon, E.E. Kheadr, J. Giasson, P. Paquin, I. Fliss, Inactivation of foodborne pathogens in milk using dynamic high pressure. *J. Food Protection.* 65 (2002) 345–352.

[8] P. Pathanibul, T.M. Taylor, P. M. Davidson, F. Harte, Inactivation of *Escherichia coli* and *Listeria innocua* in apple and carrot juices using high pressure homogenization and nisin. *Int. J. Food Microbio.* 129 (2009) 316–320.

[9] F. Sampedro, M. Rodrigo, A. Martinez, D. Rodrigo, G. Barbosa-Cánovas, Quality and safety aspects of PEF application in milk and milk products. *Critical Rev. Food Sci. Nutrition.* 45 (2005) 25–47.

[10] G.V. Barbosa-Canovas, M.M. Gongora-Nieto, U.R. Pothakamury, B. Swanson, Preservation of Foods with Pulsed Electric Fields. Academic Press, San Diego (1999).

[11] E. Ugarte-Romero, H. Feng, S.E. Martin, K.R. Cadwallader, S.J. Robinson, Inactivation of Escherichia coli with power ultrasound in apple cider, *J. Food Sci.* 71 (2006) 102–108.

[12] P. Piyasena, E. Mohareb, R.C. McKellar, Inactivation of microbes using ultrasound, *Int. J. Food Microbiol.* 87 (2003) 207–216.

[13] S.Z. Salleh-Mack, J.S. Roberts, Ultrasound pasteurization: the effects of temperature, soluble solids, organic acids and pH on the inactivation of Escherichia coli ATCC 25922, *Ultrason. Sonochem.* 14 (2007) 323–329.

[14] A.R. Baumann, S.E. Martin, H. Feng, Power ultrasound treatment of Listeria monocytogenes in apple cider, *J. Food Prot.* 68 (2005) 2333–2340.

[15] M. Valero, N. Recrosio, D. Saura, N. Munoz, N. Martıc, V. Lizama, Effects of ultrasonic treatments in orange juice processing, *J. Food Eng.* 80 (2007) 509–516.

[16] B.K. Tiwari, C.P. O'Donnell, K. Muthukumarappan, P.J. Cullen, Effect of ultrasound processing on the quality and nutritional properties of fruit juices, *Stewart Posthar.* 4 (2008) 1–6.

[17] L.H. Cheng, C.Y. Soh, S.C. Liew, F.F. Teh, Effects of sonication and carbonation on guava juice quality, *Food Chem.* 104 (2007) 1396–1401.

[18] B.K. Tiwari, C.P. O'Donnell, A. Patras, N.P. Brunton, P.J. Cullen, Effect of ozone processing on anthocyanins and ascorbic acid degradation of strawberry juice, *Food Chem.* 113 (2009) 1119–1126.

[19] V. Jiranek, P. Grbin, A. Yap, M. Barnes, D. Bates, High power ultrasonics as a novel tool offering new opportunities for managing wine microbiology, *Biotechnol. Lett.* 30 (2008) 1–6.

[20] C.D. Char, E. Mitilinaki, S.N. Guerrero, S.M. Alzamora, Use of high-intensity ultrasound and UV-C light to inactivate some microorganisms in fruit juices, *Food Bioprocess Technol.* 3 (2010) 797–803.

[21] I. Dubrovic, Z. Herceg, A.R. Jambrak, M. Badanjak, V. Dragovic-Uzelac, Effect of high intensity ultrasound and pasteurization on anthocyanin content in strawberry juice, *Food Technol. Biotechnol.* 49 (2011) 196–204.

[22] A. Vercet, C. Sanchez, J. Burgos, L. Montanes, P.L. Buesa, The effects of manothermosonication on tomato pectic enzymes and tomato paste rheological properties, *J. Food Eng.* 53 (2002) 273–278..

[23] T. Huang, C. Xu, K. Walker, P. West, S. Zhang, J. Weese, Decontamination efficacy of combined chlorine dioxide with ultrasonication on apples and lettuce, *J. Food Sci.* 71 (2006) 134–139.

[24] A. Lopez-Malo, E. Palou, M. Jimenez-Fernandez, S.M. Alzamora, S. Guerrero, Multifactorial fungal inactivation combining thermosonication and antimicrobials, *J. Food Eng.* 67 (2005) 87–93.

[25] C.P. O'Donnell, B.K. Tiwari, P. Bourke, P.J. Cullen, Effect of ultrasonic processing on food enzymes of industrial importance, *Trends Food Sci. Technol.* 21 (2010) 358–367.

[26] N.C. Cansino, G.P. Carrera, Q.Z. Rojas, L.D. Olivares, E.A. Garcia, E.R. Moreno, Ultrasound processing on green cactus pear (Opuntia ficus Indica) juice: physical, microbiological and antioxidant properties, *J. Food Process. Technol.* 4 (2013) 1–6.

[27] K.S. Suslick, Homogenous sonochemistry, in: K.S. Suslick (Ed.), Chapter in Ultrasound: Its Chemical, Physical and Biological Effects, VCH Publishers, New York, 1988.

[28] L. Leistner, Basic aspects of food preservation by hurdle technology, *Int. J. Food Microbiol.* 55 (2000) 181–186.

[29] M. Keyser, I.A. Muller, F.P. Cilliers, W. Nel, P.A. Gouws, Ultraviolet radiation as a non-thermal treatment for the inactivation of microorganisms in fruit juice, *Innov. Food Sci. Emerg. Technol.* 9 (2008) 348–354.

[30] T. Koutchma, Advances in ultraviolet light technology for non-thermal processing of liquid foods, *Food Bioprocess Technol.* 2 (2009) 138–155.

[31] J.A. Guerrero-Beltran, G.V. Barbosa-Canovas, Inactivation of Saccharomyces cerevisiae and polyphenoloxidase in mango nectar treated with UV light, *J. Food Prot.* 69 (2006) 362–368.

[32] M.O. Ngadi, J.P. Smith, B. Cayouette, Kinetics of ultraviolet light inactivation of Escherichia coli O157:H7 in liquid foods, *J. Sci. Food Agric.* 83 (2003) 1551–1555.

[33] J.A. Guerrero-Beltran, G.V. Barbosa-Canovas, Review: advantages and limitations on processing foods by UV light, *Food Sci. Technol. Int.* 10 (2004) 137–147.

[34] D.W. Donahue, N. Canitez, A.A. Bushway, UV inactivation of E. coli O157:H7 in apple juice: quality, sensory and shelf-life analysis, *J. Food Process* 28 (2004) 368–387.

[35] M. Zenker, V. Heinz, D. Knorr, Application of ultrasound-assisted thermal processing for preservation and quality retention of liquid foods, *J. Food Protect. 66* (2013) 1642–1649.

[36] E. Ugarte-Romero, H. Feng, S.E. Martin, Inactivation of Shigella boydii 18IDPH and Listeria monocytogenes Scott A with power ultrasound at different acoustic energy densities and temperatures, *J. Food Sci.* 72 (2007) 103–107.

[37] M.R. Corbo, A. Bevilacqua, D. Campaniello, D. D'Amato, B. Speranza, M. Sinigaglia, Prolonging microbial shelf life of foods through the use of natural compounds and non-thermal approaches – a review, *Food Sci. Technol. Int.* 44 (2009) 223–241.

[38] C. Franz, I. Specht, G.S. Cho, V. Graef, M.R. Stahl, UV-C-inactivation of microorganisms in naturally cloudy apple juice using novel inactivation equipment based on Dean vortex technology, *Food Control* 20 (2009) 1103–1107.

[39] L. Manzocco, B. Quarta, A. Dri, Polyphenoloxidase inactivation by light exposure in model systems and apple derivatives, *Innov. Food Sci. Emerg.* 10 (2009) 506–511.

[40] I.M. Caminiti, I. Palgan, A. Munoz, F. Noci, P. Whyte, D.J. Morgan, D.A. Cronin, J.G. Lyng, The effect of ultraviolet light on microbial inactivation and quality attributes of apple juice, *Food Bioprocess Technol.* 5 (2012) 680–686.

[41] J.M. Oteiza, L. Giannuzzi, N. Zaritzky, Ultraviolet treatment of orange juice to inactivate E. coli O157:H7 as affected by native microflora, *Food Bioprocess Technol.* 3 (2010) 603–614.

[42] V. Falguera, J. Pagan, A. Ibarz, Effect of UV irradiation on enzymatic activities and physicochemical properties of apple juices from different varieties LWT, *Food Sci. Technol.* 44 (2011) 115–119.

[43] M. Orlowska, T. Koutchma, M. Grapperhaus, J. Gallagher, R. Schaefer, C. Defelice, Continuous and pulsed ultraviolet light for nonthermal treatment of liquid foods. Part 1: Effects on quality of fructose solution, apple juice, and milk, *Food Bioprocess Technol.* 6 (2013) 1580–1592.

[44] M. Povey, T. Mason, Ultrasound in food processing. (1998) Blackie Academic and Professional, London, UK.

[45] J. Capelo-Martinez, C. Lodeiro, H.S. Santos, The power ultrasound In Ultrasound In chemistry: analytical applications. (2009) Wiley-VCH, Germany.

[46] P. Khandpur, P. R. Gogate. Evaluation of ultrasound based sterilization approaches in terms of shelf life and quality parameters of fruit and vegetable juices, *Ultrason. Sonochem.* 29 (2016) 337–353.

[47] E. Caron, G. Cheurefils, B. Barbeau, P. Payment, M. Prevost, Impact of microparticles on UV disinfection of indigenous aerobic spores. *Water Res.* 41 (2007) 4546–4556.

[48] R. Mawson, M. Raut, G. Ripoll, P. Swiergon, T. Singh, K. Knoerzer, P. Juliano, Production of Particulates From Transducer Erosion: Implications on Food Safety, Analysis Report, CSIRO Division of Animal, Food and Health Sciences, Australia, 2014.

[49] D. Bates, A. Patist, Ultrasonic innovations in the food industry: from the laboratory to commercial production. *Innovative Food Sci. Emerging Technol.* 9 (2008) 147–154.

[50] S. Muthukumaran, S.E. Kentish, G.W. Stevens, M. Ashokkumar, Application of ultrasound in membrane separation processes: A review. *Rev. Chem. Eng*. 22 (2006) 155–194.

[51] Y. Zhou, J. Shi, Effects of extracellular calcium on cell membrane resealing in sonoporation. *J. Controlled Release* 126 (2006) 34–43.

[52] C.E. Leadley, A. Williams Pulsed electric field processing, power ultrasound and other emerging technologies. In: Brennan JG, editor. Food Processing Handbook (2006) Wiley- VCH, Germany, p.214–236.

[53] E.P. Black, T. Huppertz, G.F. Fitzgerald, A.L. Kelly, Baroprotection of vegetative bacteria by milk constituents: A study of *Listeria innocua*. *Int. Dairy J.* 17 (2007) 104–110.

[54] W.A. Moats, R. Dabbah, M.V. Edwards, Survival of *Salmonella anatum* heated in various media. *App. Microbio.* 21 (1971) 476–481.

[55] I. Alvarez, P. Manas, F.J. Sala, S. Condon, Inactivation of *Salmonella enteric* serovar Enteritidis by ultrasonic waves under pressure at different water activities. *App. Env. Microbio.* 69 (2003) 668–672.

[56] S.Z. Mack, S.J. Roberts, Ultrasound pasteurization: The effects of temperature, soluble solids, organic acids and pH on the inactivation of *Escherichia coli* ATCC 25922. *Ultrason. Sonochem.* 14 (2007) 323–329.

[57] S. Ferrante, S. Guerrero, S.M. Alzamora, Combined use of ultrasound and natural antimicrobials to inactivate Listeria monocytogenes in orange juice. *J. Food Protec.,* 70 (2007) 1850–1856.

[58] R.C. Mckellar, E. Mohareb, P. Piyasena, Inactivation of microbes using ultrasound: a review. *Int. J. Food Microbio.* 87 (2003) 207–216.

[59] H. Alliger, Ultrasonic Disruption. American Laboratory 10 (1975) 75–85.

[60] G. Barobosa-Canovas, H. Feng, J. Weiss, Ultrasound technologies for food and bio processing, (2011) Springer, USA.

[61] S. Drakopoulou, S. Terzakis, M.S. Fountoulakis, D. Mantzavinos, T. Manios, Ultrasound induced inactivation of gram-negative and gram-positive bacteria in secondary treated municipal wastewater. *Ultrason. Sonochem.* 16 (2009) 629–634.

[62] W.C. Frazier, C. D. Westhoff, Food Microbiology. 4th ed. (1988) McGraw- Hill, Inc, New York.

[63] P. Manas, R. Pagan, Microbial inactivation by new technologies of food preservation. *J. App. Microbio.* 98 (2005) 1387-1399.

[64] L. Ciccolini, P. Taillandier, A.M. Wilhem, H. Delmas, P. Strehaiano, Low frequency thermo-ultrasonication of *Saccharomyces Cerevisiae*

suspensions: effect of temperature and of ultrasonic power. *Chem. Eng. J.* 65 (1997) 145-149.

[65] R. Bhat, N.S.B.C. Kamaruddin, L. Min-Tze, A.A. Karim, Sonication improves kasturi lime (Citrus microcarpa) juice quality, *Ultrason. Sonochem.* 18 (2011) 1295–1300.

[66] D. Knorr, M. Zenker, V. Heinz, D. Lee, Applications and potential of ultrasonics in food processing. *Trends Food Sci. Tech.* 15 (2004) 261-264.

[67] P. Lopez, A. Vercet, A.C. Sanchez, J. Burgos, Inactivation of tomato pectin enzymes by manothermosonication. *Zeitschrift fur Lebensmittel-Untersuchung und –Forschung.* 207 (1998) 249–252.

[68] A.R. Sams, R. Feria, Microbial effects of ultrasonication of broiler drumstick skin. *J. Food Sci.* 56 (1991) 247–248.

[69] P. Fellows, Food Processing Technology: Principles and Practice, 2nd ed. (2000) *CRC Press,* New York.

[70] P. Butz, B. Tauscher, Emerging technologies: chemical aspects. *Food Res. Int.* 35 (2002) 279–284.

[71] A. Baumann, S.E. Martin, H. Feng, Power ultrasound treatment of Listeria monocytogenes in apple cider. *J. Food Protect.* 68 (2005) 2333–40.

[72] M. Valero, N. Recrosio, D. Saura, N. Munoz, N. Marti, V. Lizama, Effects of ultrasonic treatments in orange juice processing. *J. Food Eng.* 80 (2007) 509–516.

[73] S. Patil, P. Bourke, B. Kelly, J.M. Frías, P.J. Cullen, The effects of acid adaptation on Escherichia coli inactivation using power ultrasound. Inn*ov. Food Sci. Emerging Tech.* 10 (2009) 486–490.

[74] V.M. Gomez-Lopez, L. Orsolani, A. Martinez-Yepez, M.S. Tapia, Microbilogical and sensory quality of sonicated calcium-added orange juice. *LWT- Food Sci. Tech.* 43 (2010) 808–813.

[75] L.H. Cheng, C.Y. Soh, S.C. Liew, F.F. Teh, Effects of sonication and carbonation on guava juice quality. *Food Chem.* 104 (2007) 1396−1401.

[76] A.O. Adekunte, B.K. Tiwari, P.J. Cullen, A.G.M. Scannell, C.P. O'Donnell, Effect of sonication on colour, ascorbic acid and yeast inactivation in tomato juice. *Food Chem.* 122 (2010) 500–507.

[77] P. Quek, J. Hu, Indicators for photoreactivation and dark repair studies following ultraviolet disinfection. *J. Ind. Microbio. Biotech.* 35 (2008) 533–541.

[78] M. Begum, D. Hocking-Ailsa, D. Miskelly, Inactivation of food spoilage fungi by ultra violet (UVC) irradiation. *Int. J. Food Microbio.* 129 (2009) 74–77.
[79] T. Bintsis, E.L. Tzanetaki, R.K. Robinson, Existing and potential applications of ultraviolet light in the food industry – a critical review. *J. Sci. Food Agriculture.* 80 (2000) 637–645.
[80] M.T.T. Tran, M. Farid, Ultraviolet treatment of orange juice. *Innov. Food Sci. Emerging Tech.* 5 (2004) 495–502.
[81] L.H. Choi, S. S. Nielsen, The effects of thermal and nonthermal processing methods on apple cider quality and consumer acceptability. *J. Food Quality* 28 (2005) 13–29.
[82] D.E. Hanes, P. A. Orlandi, D. H. Burr, M. D. Miliotis, M. G. Robi, J. W. Bier, G. J. Jackson, M. J. Arrowood, J. J. Churey, R. W. Worobo. Inactivation of Cryptosporidium parvum oocysts in fresh apple cider using ultraviolet irradiation. *App. Env. Microbio.* 68 (2002) 4168–72.
[83] N. Basaran, A. Quintero-Ramos, M.M. Moake, J.J. Churey and R.W. Worobo, Influence of apple cultivars on inactivation of different strains of Escherichia coli O157:H7 in apple cider by UV irradiation. *App. Env. Microbio.* 70 (2004) 6061–6065.
[84] E. Caron, G. Chevrefils Jr, B. Barbeau, P. Payment, M Prevost, Impact of microparticles on UV disinfection of indigenous aerobic spores. *Water Res.* 41 (2007) 4546-4556.
[85] E.G. Murakami, L. Jackson, K. Madsen, B. Schickedanz, Factors affecting the ultraviolet inactivation of Escherichia coli K12 in apple juice and a model system. *J. Food Proc. Eng.* 29 (2006) 53–71.
[86] A. Hassen, M. Mahrouk, H. Ouzari, M. Cherif, A. Boudabous, J.J. Damelincourt, UV disinfection of treated wastewater in a large scale pilot plant and inactivation of selected bacteria in a laboratory UV device. *Biores. Tech.* 74 (2000)141–150.
[87] T. Koutchma, B. Parisi, E. Patazca, Validation of UV coiled tube reactor for fresh juices. *J. Env. Eng. Sci.* 6 (2007) 319–328.
[88] J.A. Parker, J.L. Darby, Particle-Associated Coliform in Secondary Effluents: Shielding from Ultraviolet Light Disinfection. *Water Env. Res.* 67 (1995) 1065-1072.
[89] H. P. Vasantha Rupasinghe, L. J. Yu, Emerging Preservation Methods for Fruit Juices and Beverages, in Food Additive Ed. by Prof. Yehia El-Samragy, (2012) ISBN: 978-953-51-0067-6.

[90] M. Zenker, V. Hienz, D. Knorr, Application of ultrasound-assisted thermal processing for preservation and quality retention of liquid foods. *J. Food Protec.* 66 (2003) 1642−1649.

[91] S.L. Rodgers, T.E. Ryser, Reduction of microbial pathogens during apple cider production using sodium hypochlorite, copper ion, and sonication. *J. Food Protec.* 67 (2004) 767–771.

[92] A. López-Malo, E. Palou, Ultraviolet light and food preservation, in Novel Food Processing Technologies (2005) eds G. V. Barbosa-Cánovas, M. S. Tapia, M. P. Cano *CRC Press, Madrid, Spain.* 464–484.

[93] D. D'Amico, T. Silk, J. Wu, M. Guo, Inactivation of microorganisms in milk and apple cider treated with ultrasound. *J. Food Protec.* 69 (2006) 556–563.

[94] J. Kuldiloke, M.N. Eshtiaghi, Application of non-thermal processing for preservation of orange juice. *KMITL Sci. Tech. J.* 8 (2008) 64–74.

[95] Y. Yuan, Y. Hu, T. Yue, T. Chen, Y. M. Lo, Effect of ultrasonic treatments on thermoacidophilic *Alicyclobacillus acidoterrestris* in apple juice. *J. Food Proc. Preserv.* 33 (2009) 370–383.

[96] S. Utkun, B. Kunduhoglu, Combined effect of ultrasound and selected essential oil constituents on Escherichia coli O157:H7 and Listeria monocytogenes reduction in orange juice. *African J. Microbio. Res.* 6 (2012), 6642–6652.

[97] N. Dizadji, P. Entezar, A. Afsari, Experimental Study of Kiwi Juice under Sonication and Carbonation. World Academy of Science, Engineering and Technology. 61 (2012) 1060.

[98] J. Gao, H. P. Vasantha Rupasinghe, Nutritional, Physicochemical and Microbial Quality of Ultrasound-Treated Apple-Carrot Juice Blends. *Food Nutrition Sci.* 3 (2012) 212-218

[99] Q.Y. Zafra-Rojas, N. Cruz-Cansino, E. Ramírez-Moreno, L. Delgado-Olivares, J. Villanueva-Sánchez, E. Alanís-García, Effects of ultrasound treatment in purple cactus pear (Opuntia ficus-indica) juice. *Ultrason Sonochem* 20 (2013) 1283-1288.

[100] H.R. Alighourchi, M. Barzegar, M. A. Sahari, S. Abbasi. Effect of sonication on anthocyanins, total phenolic content, and antioxidant capacity of pomegranate juices. *Int. Food Res. J.* 20 (2013) 1703–09.

[101] K. Tandon, R.W. Worobo, J.J. Churey, O.I. Padilla-Zakour, Storage quality of pasteurized and UV treated apple cider. *J. Food Proc. Preserv.* 27 (2003) 21-35.

[102] G. Duthie, A. Crozier, Plant-derived phenolic antioxidants, *Current Opinion Lipidology.* 11 (2000) 43-47.

[103] A. Burt, Essential oils: Their antibacterial properties and potential applications in foods: A review. *Int. J. Food Microbio.* 94 (2004) 223-253.

[104] A.A. Ortuno, P. Baidez, C.M. Gomez, I. Arcas, G.A. Porras, A.J. Del Rio, Citrus paradise and Citrus sinensis flavonoids: Their influence in the defense mechanism against Penicillium digitatum, *Food Chem.* 98 (2006) 351-358.

[105] L. Elisa, A. Ignacio, R. Javier, Improving the pressing extraction of polyphenols of orange peel by pulsed electric fields. *Innov. Food Sci. Emerging Technol.* 17(2013) 79-84.

[106] Y. Chun-Lin, D. De-Hui, H. Wei-Lian, Antimicrobial and antioxidant activities of the essential oil from onion (Allium cepa L.). *Food Control.* 30 (2013) 48-53.

In: Applications of Ultrasound in the Beverage… ISBN: 978-1-63485-069-8
Editor: Juan Francisco García Martín

Chapter 5

ULTRASOUND FOR ACCELERATING THE WINE AGEING PROCESS FROM PHYSICOCHEMICAL POINT OF VIEW

Juan Francisco García[1*], Qing-An Zhang[2] and Chao-Hui Feng[3]

[1]Department of Chemical Engineering, University of Malaga, Campus Teatinos, Malaga, Spain

[2]School of Food Engineering and Nutrition Science, Shaanxi Normal University, Xi'an, China

[3]College of Food Science, Sichuan Agricultural University, Ya'an, Sichuan, China

ABSTRACT

Due to the loss of competitiveness of wine producers during the last decade, researchers and wine producers have recently paid much attention to ultrasound as a novel technique for accelerating the wine ageing process. This is because if ultrasound were able to bring about the same effects that many years of natural ageing, this would constitute a major breakthrough for the winemaking industry, and would extremely improve the competitiveness of wine producers using ultrasound technology in the face of the increasing global competition for wine sale areas. However,

* Corresponding Author Email: jfgarmar@uma.es.

ultrasound should not be regarded as a novel technology in winemaking since its effects on wine were extensively assayed from the late 30's until the early 60's of the last century. Results from both the 20th century and the last decade are quite similar, and the actual effects of ultrasound remain to be demonstrated. In this chapter, we reviewed what have we have found so far in relation to the effects of ultrasound on some physicochemical properties (pH, acidity, phenolic compounds ...) and on free radical generation in a red wine. Furthermore, we propose new alternatives for ultrasound to assist the wine ageing process. Finally, we explain a method for online monitoring the effects of ultrasound on wine based on electrical conductivity measures.

Keywords: online monitoring, ultrasound, wine ageing, winemaking

EARLY ASSAYS OF ULTRASOUND ON WINE

Wine producers are constantly striving to achieve a stable product with an extended period of peak taste and bouquet. Innovative winemaking and ageing technologies are currently under development due to the disadvantages of the traditional ageing in barrels such as lengthy time needed, high cost, etc. (Tao et al., 2014a). At present, the application of ultrasound to wine is regarded as one of the most promising and widely studied techniques. However, ultrasound cannot be properly considered as a novel technology for winemaking. In 1937, ultrasonic waves were applied for the first time to produce thermodynamic changes in fermented and distilled alcoholic beverages to emulate those changes that occur in a natural aged beverage (Bachmann and Willkins, 1937). During the following decade, ultrasound treatment of wines and spirits was reported several times to produce quality improvements and therefore to constitute a potential quick ageing technique until the work by Singleton and Draper about the application of ultrasound to wine on the early 60's of the last century. These authors (Singleton and Draper, 1963) assessed the effects of ultrasound on 5 kind of wine (white sweet, red sweet, red dry, white dry, and shermat) using different gas purging (nitrogen, air, hydrogen, oxygen, and carbon dioxide) and concluded that although 1-hour 90-kHz ultrasound treatment did not significantly improve the wine quality, some parameters did change. To be specific, it was found a slightly increase of the volatile acidity and the visible absorption of red wine. Tannin concentration also increased when working under nitrogen, hydrogen and carbon dioxide atmosphere, but it generally decreased when the ultrasound treatments were performed with air and oxygen. On the contrary, these authors (Singleton and Draper, 1963) found that changes in volatile esters in wine tended to be negative after ultrasound treatment, which could be related to a possible degassing effect (volatile compounds removing) of ultrasound (García Martín and Sun, 2013). Regardless the gas (nitrogen, air, hydrogen, oxygen or carbon dioxide), Singleton and Draper (1963) detected the same flavor in the 5 kind of wine assayed after ultrasound treatment. Their panelists described it as a kind of "scorched" flavor. The authors then suggested that ultrasound decreased grape aroma in wine, which could be reasonable since most of aromas come from volatile compounds and may be easily lost by the degassing effect of ultrasound. Finally, these early assays under different atmospheres showed that oxidation was not increased in the presence of air or nitrogen, but was clearly accelerated when oxygen was introduced (Singleton and Draper, 1963). Since then there has been a lack of researches about the effects of ultrasound on wine during the remaining 20th century, probably due to the negative, erratic results provided by Singleton and Draper (1963).

WHY ULTRASOUND AGAIN?

At the beginning of the 21st century, several works, mainly patents, had attempted to literally ultrasonically accelerate the ageing of several alcoholic beverages (Tyler III and Bailey, 2003; Leonhardt and Morabito, 2007). Besides the advantage of shortening the ageing time, these works also reported that ultrasound can keep the quality of wine at a high level for a long time. Although wines may loss the aroma and fermentation bouquet during ageing, they also gain a bottled-aged bouquet (Jackson, 1994). When wines are aged for a certain period, they reach their "peak" and have the most pleasant taste and softening of tannins, most complicate aged bouquet (Robinson, 2006). After that, the wine quality starts to irreversibly decline. Therefore, wines should be consumed when they reach their respective "peaks." Leonhardt and Morabito (2007) claimed that wines can reach their "peak" much quicker and wine quality can remain stable at the "peak" level for a much longer time when using ultrasound (Figure 1), thus extending the shelf life of wine. To the best of our knowledge, this affirmation remains to be demonstrated.

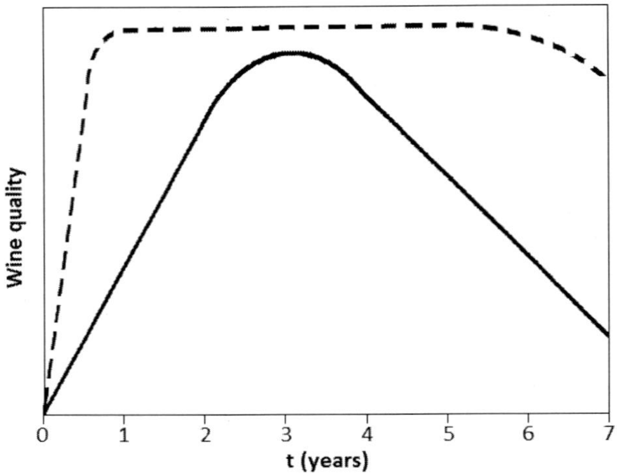

Figure 1. Comparison between the quality evolution of natural aged wines (continuous line) and the expected quality of ultrasonic aged wines (dashed line).

At that time, the competitiveness of European wine producers was already under serious threat due to New World wines that had been gaining an increasing share of the European market during the previous years. Chilean, South African, New Zealand and Australian wines are generally less costly than the cheapest European wines which are usually produced using traditional production methods. What is more, both Australia and New Zealand were achieving record grape harvests causing a worldwide wine glut. For example, New Zealand wine production rose 20% in the 2006-2007 harvesting. A large quantity of this production was destined for the export market, thus starting competition in European wine sale areas. In spite of the fact that European wines traditionally feature among the best in the world, and that more than 60% of the world's wine still came from Europe, wine producers were faced with mounting global competition. As a result, the European Union paid more than 180 million euros in 2007 for the distillation of European wines which could not be sold and around 130 million euros in 2008 for surplus wine from European wineries. It is worth noting that there are an estimated 1.5 million wine orchards, covering 3.4 million hectares, i.e., 2% of EU agricultural area.

Within this context, one could think that an ultrasound equipment that really work for assisting and accelerating the wine ageing process would represent a major breakthrough for wine producers. Based on this threat to European wine producers and the promising results provided by the aforementioned patents about the potential of ultrasound treatment to bring about the same effect as natural ageing on wine, a major "Research for SMEs" project entitled "Novel method for assisting and accelerating the aging process of wine" (ULTRAFINEWINE) was funded by the European Commission's Seventh Framework Programme (FP7) and had the backing of six companies, ranging from wine producers to equipment manufacturers, and several Research Institutions from across Europe. The main goal of this project was to design a prototype ultrasound device, which in turn would be scaled up to integrate it into existing wine fermentation vats from the participating SMEs. If the production wine resembling many years of natural ageing were achieved, the participating SMEs would improve the competitiveness in the face of such increasing global competition. This project represents the major effort to date to ultrasonically age wines and validate its usefulness at industrial scale.

In spite of the interesting approach of this project, some issues must be addressed. First at all, the proposed ultrasound-assisted ageing procedure aimed to substitute the traditional wine ageing in barrels, which is the most common method in winemaking process. Although this traditional technology has several drawbacks, such as the long time that this process requires and the high cost of barrels production, ageing in barrels is still the most widely used and recognized technique. Oak wood has been used to build wine barrels for over 2000 years (Jackson, 1994). Oak barrels can benefit wines in two different aspects. On one hand, astringency-related phenolic compounds and oak-responsible aromatic compounds are transferred to wine during ageing (Tao et al., 2014a). This would not occur during the ultrasound-assisted ageing. On the other hand, oxygen is introduced at various stages during winemaking. Gentle oxygenation can contribute to enhance wine quality and increasing exposure of wine to oxygen can accelerate the rate at which red wine matures, while excessive oxygenation is detrimental to wine quality and hence can shortens the peak of the wine (García Martín and Sun, 2013). Therefore, small quantities of dissolved oxygen during ageing are required. The semi-permeable wood walls allow slow atmospheric oxygen permeation through the barrel which provokes the gently oxidation of certain compounds, leading to a reduction of astringency and slight changes in color (Bozalongo et al., 2007). In a closed ultrasound reactor, this would not occur either. The degassing effect of ultrasound could also lead to unexpected results in oxidation, condensation and polymerization reactions of phenolic compounds in wine (García Martín and Sun, 2013).

There are some actual alternatives to overcome these two issues. For the former, small pieces of wood, usually known as oak chips, could be added into wine vats in order to produce a woody aroma and taste on wine (Morales et al., 2004). In this sense, a first study about the release of oak-related compounds from oak chips under ultrasound treatment can be already found in literature (Tao et al., 2014b). These authors experimentally studied and mathematically modeled the release kinetics of total phenolics from oak chips into a model wine (12% (v/v) aqueous ethanol solution, acidulated to pH 3.5 with tartaric acid) at various conditions (ultrasound energy density, ultrasound exposure time, and bath temperature), and found that ultrasound exposure time was the operational parameter with the greatest influence on the increasing the release of oak-related compounds. For the latter alternative, micro-oxygenation could be employed to introduce small, measured amounts of oxygen into wines, thus simulating the mild oxygenation of wine during barrel ageing (Guerrero et al., 2010). Therefore, the combination of ultrasound with wood fragments or microoxygenation could have a great potential in winemaking.

Apart from ageing in barrels (maturation), bottling is another important stage (in fact, the second stage) for wine ageing. Wines are usually packaged in glass bottles with air-tight cap to avoid direct contact with air. Bottling has been called "reductive ageing" because oxygen exposure is almost nil (Jackson, 1994). The dissolved oxygen in bottled wines depends on the type of closure and the materials of bottle. There are several types of closure, e.g., cork stopper, synthetic closures, screw caps. The cork system is widely used to seal bottles and limit the permeation of atmospheric oxygen into bottles (Tao et al., 2004a). In our opinion, the acceleration of the reductive ageing is the most likely effect of ultrasound, without combining with other methods, on wine.

Lastly, the presence of lees, generally during ageing in barrels, can improve the organoleptic characteristic (decreasing the astringency and bitterness of tannins and softening wood flavor) of some wines. Yeast autolysis can enrich wine by compounds released from cytoplasm and cell wall, such as polysaccharides, mannoproteins, and other low molecular weight products. The presence of polysaccharide-tannin complexes can reduce wine astringency (Del Barrio-Galán et al., 2012), while mannoproteins can prevent polymerization of monomeric anthocyanins, thus enhancing wine color stabilization (Zamora, 2002). Ultrasound has already proved its feasibility to provoke high cell disruption on yeasts from wine lees (García Martín et al., 2013). Therefore, from the point of view of the winemaking, the use of ultrasound, a low-cost technology, during ageing on lees could drastically reduce the yeast autolysis time (from months to hours), thus releasing large quantities of polysaccharides and mannoproteins in a very short time (García Martín et al., 2013). Loira et al. (2016) have recently assayed the sequential application of ageing on lees and ultrasound, without adding oak chips or performing the ultrasound experiments in oak barrels, thus limiting the potential results. These authors aged on lees at laboratory scale a red wine made of Tempranillo grapes from DO Ribera del Duero and subsequently the aged wines were subjected to ultrasound treatment for 8 weeks, four days a week, one hour per day. It was confirmed the loss of aromatic compounds after ultrasound treatment, probably due to the aforementioned degassing effect (García Martín and Sun, 2013). Unlike other studies in which a negative flavor (sometimes described as "scorched") was associated to ultrasound treatment (Singleton and Draper, 1963; Cui et al., 2012), their panelists did not detect any particular odor or flavor in the wines treated with ultrasounds, but ultrasound treatment failed to reduce the astringency and acidity of the wine, nor help to improve its body.

As it can be seen, the potential of ultrasound in winemaking is undoubtable and ultrasound should be combined with several already implemented winemaking technologies (oak chips addition, microoxygenation or ageing on lees). Nevertheless, these combinations have not been assayed to date. Researches to date are mainly focused on the effects of ultrasound alone on different kind of wines (Singleton and Draper, 1963; Masuzawa et al., 2000; Chang and Chen, 2002; Chang, 2004; Chang, 2005; Cui et al., 2012; Zhang et al., 2016a, 2016b), and the results and conclusions are uneven, so the actual effects of ultrasound on grape wine still remain to be demonstrated.

ACTUAL CHEMICAL EFFECTS OF ULTRASOUND ON RED WINE

During the last two years our research team has tried to elucidate the actual direct effects of ultrasound on red wine from a chemical point of view, as reported elsewhere (Zhang et al., 2015a, 2016a, 2016b). Using the same red wine (Cabernet Sauvignon red wine from the vintage 2012), we have assessed the influence of the main operational parameters, namely ultrasound power (120, 150, 180, 210, 240, 270 and 300 W), ultrasound frequency (45, 80 and 100 kHz), exposure time (20, 40, 60, 80 and 100 min) and bath temperature (20ºC, 30ºC, 40ºC, 50ºC and 60ºC), on red wine.

Effects on Basic Oenological Parameters

Chemical parameters such as pH, total and volatile acidity, ethanol and sulfur dioxide are generally used to define and to express wine quality (García Martín and Sun, 2013). Acidity and pH are very important parameters for the wine stability. The increase of either pH or volatile acidity can lead to the growth of undesirable microorganisms. By contrast, the pH should be not too low when malolactic fermentation is desired. The main acids in wine are tartaric and malic acids. Acetic acid, which is produced by yeasts and lactic acid bacteria during fermentation, is the main volatile acid. At high concentrations, this compound is responsible for "off-flavor," a vinegar note. The active ingredient in alcoholic beverages is ethanol. Grapes are around 80% water and 20% sugars, which are fermented to ethanol during the winemaking process. Sulfur dioxide can increase wine color by preventing the formation of insoluble complexes as a result of the oxidative polymerization of red pigments and other flavonoids in wine. By contrast, high amounts of free sulfur dioxide can decrease color intensity as SO_2 is an effective bleaching agent (Jackson, 1994).

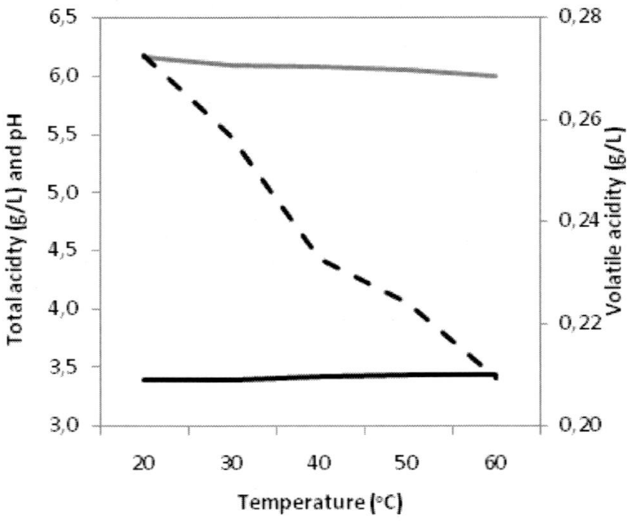

Figure 2. Total acidity (grey line), pH (black line) and volatile acidity (dashed line) in red wine after ultrasound treatment at different temperatures.

In a previous research, we found that the different assayed combinations of ultrasound power, frequency and exposure time did not exert any appreciable effects on pH and total acidity of red wine (Zhang et al., 2016a). We have now verified this fact by measuring volatile acidity (data not shown). By contrast, we have found a slight influence of temperature on volatile acidity (Figure 2) which, consequently, led to a mild decreasing in total acidity and, in turn, to a slight increasing of pH. This fact is reasonable since the higher the temperature is, the higher the volatilization rate of acetic acid. However, this is contrary to Singleton and Draper's findings, who reported a slight increase of the volatile acidity in red wine after ultrasound treatment (Singleton and Draper, 1963). In regards to sulfur dioxide, the degasification effect of ultrasound could lead to an important decrease of sulfur dioxide, mainly free SO_2, in wine. This could increase the oxidation rate of wine, since the dissolved SO_2 has antioxidant properties, preserving wines from both enzymatic and chemical oxidation (García Martín and Sun, 2013). Unfortunately, we do not have consistent data to confirm this fact, since our results about sulfur dioxide in red wine are a bit erratic (data not shown). What is more, Cui et al. (2012) did not find obvious differences on free SO_2 and total SO_2 in Italian Riesling white wine after 20-min ultrasound treatment either. Finally, ultrasound did not affect the alcohol level of the Cabernet Sauvignon red wine used (data not shown). All in all, ultrasound does not seem to exert any noticeable effect on basic oenological parameters of red wines.

Changes in Phenolic Composition

The main target of ultrasound during winemaking should be promoting and accelerating condensation reactions between anthocyanins and other phenols (mainly tannins), thus shortening the time required for wine ageing. However, the solely reference that can be found to date in literature in this regard is that of Masuzawa et al. (2000), who pointed out that ultrasound promoted the polymerization of phenolic compounds of red wines, without specifying what kind of reactions were promoted by ultrasound or the involved phenolic compounds. Generally, the total phenolic concentration of red wines falls during ageing regardless the phenolic content of the grape variety or varieties and the winemaking process. In our research (Zhang et al., 2016a) we found that the concentration of total phenolic compounds in ultrasonically treated wines was slightly lower (ranging between 1.22 and 1.29 g/L) than that of the untreated Cabernet Sauvignon red wine (1.31 g/L). The changes of total phenolic compounds were a bit erratic in all the trials performed at different ultrasound powers, ultrasound frequencies, exposure times and bath temperatures, and no trend could be found. This is in accordance with the results reported by Singleton and Draper (1963).

Principal component analysis (a technique often used to make data easy to explore and visualize) of total phenolic compounds along with basic oenological properties of wines for each assayed parameter (ultrasound power, ultrasound frequency, exposure time and bath temperature) resulted in an effective clustering of wines into two groups: the ultrasonic treated wines and the untreated wine (Zhang et al., 2016a). This confirmed that ultrasound modified the physicochemical characteristics of red wine, but did not provide information about in which way. Principal component analysis also suggested that exposure time had the greatest influence on total phenolic compounds.

Changes in Color and Chromatic Characteristics

Wine tint and intensity are a visual indicator to judge the quality of red wine and affect consumer's satisfaction, which are also related to the ageing process since color intensity of red wine decreases during ageing while tint increases (García Martín and Sun, 2013). In our research (Zhang et al., 2016a), analysis of variance of wine tint and intensity values showed no significant differences. In order to assess changes in chromatic characteristics, the best parameter is the overall colorimetric difference between two wines (ΔE^*), which takes into account the differences in the colorimetric coordinates L^* (clarity), a^* (red/green color component) and b^* (yellow/blue color component) as follow: $\Delta E^* = [(\Delta L^*)^2 + (\Delta a^*)^2 + (\Delta b^*)^2]^{1/2}$. Therefore, the overall colorimetric differences with respect the untreated wine were calculated in order to evaluate whether the changes in color promoted by ultrasound were noticeable. The theoretical limit of perception for the human eye ($\Delta E^* \geq 3$) has been suggested as an absolute color discrimination threshold for red wines (García Martín and Sun, 2013). Although no pattern was found, ΔE^* values were close to this threshold after sonication in our research (Zhang et al., 2016a). The red wine treated with 240 W ultrasound and the one treated with 100 kHz ultrasound reached a ΔE^* value of 2.8 while the red wine ultrasonically treated for 80 min reached a ΔE^* value of 2.7. In general, all the wines treated at 20°C with different ultrasound powers, frequencies and exposure times had remarkably ΔE* values. On the contrary, much lower colorimetric differences (between 1.0 and 1.5) were found in red wines treated at temperatures above 20°C. Therefore, the increasing of bath temperature reduced the effect of ultrasound in chromatics characteristics.

Free Radical Generation and Free Radical Scavenging Activity

There is a lack of papers concerning the ultrasound action on wines. Generally, free radicals are considered as the important triggering factors to initiate chemical reactions within liquids exposed to ultrasound. Besides the hydroxyl radicals produced by ultrasound in aqueous solution (Kubo et al., 2012), some other free radicals are also generated by ultrasonic waves in some organic liquids (Castellanos et al., 2001), which in turn may react with other compounds and generate new radicals thus leading to a chain of chemical reactions. For these reasons, we assessed whether ultrasound has the ability to induce free radicals in red wine, and what kind of free radicals are generated (Zhang et al., 2015a). In that paper, we firstly focused on the identification, by electron paramagnetic resonance spin trapping, of free radicals induced by low frequency ultrasound in the Cabernet Sauvignon red wine and also in a model wine, which contained 12% (v/v) ethanol, 53 mM tartaric acid and 12 mM catechin. Once the free radicals were identified, we assessed the effect of the main operational parameters aforementioned (ultrasound power, ultrasound frequency, ultrasound exposure time and bath temperature) on the intensity of the free radical spin adducts.

The use of electron paramagnetic resonance spin trapping of free radicals with 5,5-dimethyl-1-pyrroline N-oxide (DMPO) succeeded in identifying the species of free radicals generated by ultrasound irradiation in red wine and model wine. The results showed that the 1-hydroxylethyl free radical was captured in both red wine and model wine, and its concentration increased after ultrasound treatment, which demonstrates that ultrasound did trigger the generation of 1-hydroxylethyl free radicals into wine. This free radical is sufficiently stable to be trapped using nitrone spin traps as 5,5-dimethyl-l-pyrrolin N-oxide and was therefore quantified by measuring the intensity of the electron paramagnetic resonance spectrum corresponding to the spin adduct. By contrast, the electron paramagnetic resonance spectrum of hydroxyl free radical was not detected in either red wine or model wine. The reason was that ethanol, as a hydroxyl free radical scavenger, has the ability to react with hydroxyl free radicals leading to the formation of 1-hydroxylethyl free radicals according to the literature available (Makino et al., 1983; Kreitman et al., 2013), i.e., ethanol is a substrate for the hydroxyl free radical and therefore competes with the spin trap (Makino et al., 1983). Furthermore, the latter free radical newly formed is more stable than the former ones. As a result, the concentration of hydroxyl free radical in solutions containing ethanol is almost nil, which was confirmed by our electron paramagnetic resonance spectra (Zhang et al., 2015a). On the other hand, we found that the presence of catechin in the model wine exhibited a certain scavenging activity on the 1-hydroxylethyl free radical during ultrasound exposure, which was not found when using the same model wine without catechin. With regard to the operational parameters of ultrasound in model wine, the increase of ultrasound power, frequency and exposure time resulted in an increase in the intensity of DMPO/1-hydroxylethyl free radical spin adducts. The increase of temperature form 20 to 50ºC also promoted the intensity of the DMPO/1-hydroxylethyl free radical spin adducts, followed by a decrease at the higher assayed temperature (60ºC), which might be attributed to the chemical instability of DMPO/1-hydroxylethyl free radical at high temperatures (Buettner, 1987). The higher temperature of 60ºC might cause the increase of the degradation rate of DMPO/1-hydroxylethyl free radical (Kubo et al., 2012), and when its degradation rate is greater than that of its formation, a falling trend could be observed.

Figures 3-5 show the detection of the 1-hydroxylethyl free radical formation in red wine after sonication (Zhang et al., 2105a). The hyperfine coupling constants of the spectrum (a_N = 15.60 G, a_H = 2.25 G) were in agreement with the values for the DMPO spin adducts formed from the 1-hydroxyethyl radical in red wine previously reported by Elias et al. (2009). Figure 4 confirms that this kind of free radical does exist in red wine while Figure 5 demonstrates that ultrasound treatment contributes to the formation of the 1-hydroxylethyl free radical directly or indirectly.

On the other hand, we have also assessed the 1,1-diphenyl-2-picryl-hydrazyl (DPPH) free radical scavenging activity induced by ultrasound in the same Cabernet Sauvignon red wine (Zhang et al., 2016a). We found that ultrasound frequency did affect the scavenging activity of DPPH, which decreased to 93.02% when increasing the ultrasound frequency from 45 to 80 kHz. By contrast, the scavenging activity increased using a higher ultrasound frequency (100 kHz) and no explanation was found. Statistical results showed that the use of different ultrasound powers (ranging from 240 to 300 W), exposure times (from 20 to 100 min) and bath temperatures (between 20 and 60ºC) did not influence the DPPH free radical scavenging activity. In red wines, phenolic compounds are responsible for DPPH free radical scavenging activity. Several authors (Fernández-Pachón et al., 2004; Lucena et al., 2010; Sánchez-Moreno et al., 1999) have pointed out that the total phenolic compounds concentration is closely related to antioxidant activity. The correlation between TPC and DPPH-free radical scavenging activity was then performed. The coefficient correlation was found to be 0.539, which is in accordance with the findings of these authors, hence confirming the antioxidant activity of wine phenolic compounds (Zhang et al., 2016a). What is more, statistically significant changes in the DPPH scavenging activity were related to changes in the concentration of total phenolic compounds after ultrasound treatment.

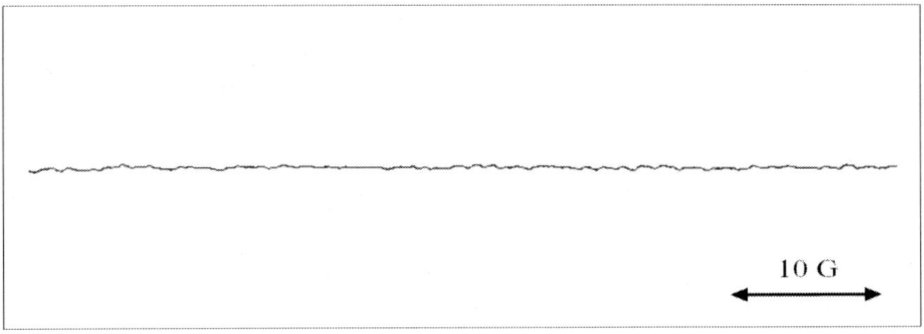

Figure 3. Electron paramagnetic resonance spectrum of red wine without DMPO.

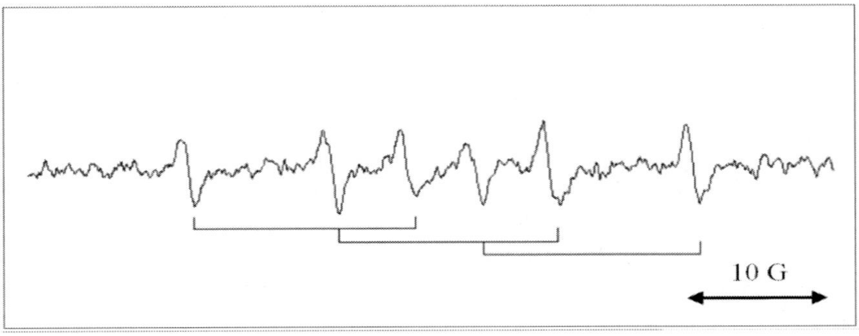

Figure 4. Electron paramagnetic resonance spectrum of red wine after 500 mM DMPO addition.

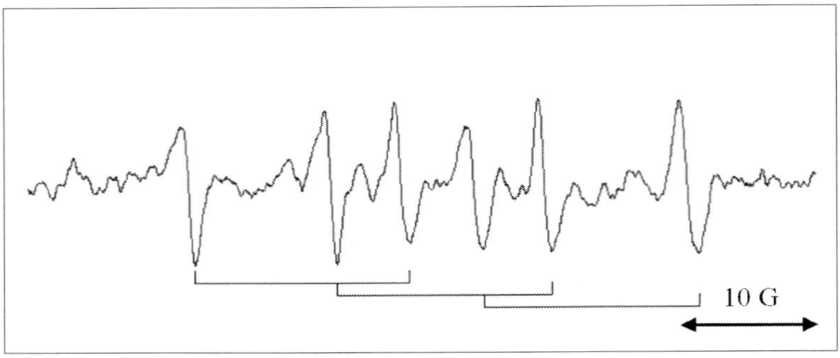

Figure 5. Electron paramagnetic resonance spectrum of red wine with 500 mM DMPO after 5 min ultrasound exposure.

Monitoring the Effects of Ultrasound on Wine by Electrical Conductivity

Electrical conductivity is an electrical phenomenon which is defined as the measure of a material's ability to accommodate the transport of an electric charge, and it is calculated as the ratio of the current density to the electric field strength. The electrical conductivity of a solution is highly proportional (but not totally) to its ion concentration. However, conductivity does not always correlate directly to concentration.

This important physicochemical property of food products has attracted interest in the wine industry because its good correlation with pH and assimilable nitrogen during fermentation (Colombié et al., 2007, 2008). Wine primary and secondary fermentations are still carried out without on-line monitoring due to the lack of robust, precise and cheap sensors (Colombié et al., 2008). The assimilation of nitrogen (in particular, ammoniacal nitrogen) is usually the limiting nutrient of the fermentation kinetics, hence the interest of on-line monitoring electrical conductivity during winemaking. Besides, an estimation of the wine cold stability after fermentation (to prevent tartaric salts from precipitating out the wine when stored) based on the decrease in conductivity of wine over a period of time has been proposed as well (Zoecklein et al., 1995). Therefore, the online monitoring of electrical conductivity during winemaking could be used as a reliable and cheap sensor.

Significant changes of electrical conductivity have been observed by different authors (Jambrak et al., 2007; Aadil et al., 2013) in food exposed to ultrasound, but these changes were measured discontinuously during ultrasound treatment, so these recorded values after ultrasound treatment could differ from electrical conductivity values during sonication. The real-time evaluation of changes induced by ultrasound in a specific wine compound is difficult because of the complex composition of wine. Furthermore, most of the methods for wine characterization are destructive and therefore cannot be monitored online. For this reason, the online monitoring of wine quality by means of electrical conductivity could be highly useful to assess the actual effects of ultrasound on wine.

In our research (Zhang et al., 2016b), we assessed the effect of the aforementioned operational parameters (ultrasound power, ultrasound frequency, exposure time and temperature bath) on wine electrical conductivity, using the same Cabernet Sauvignon red wine. Interestingly, we found that the electrical conductivity value measured discontinuously at a given time was very different from the real-time monitored value at the same given time during ultrasound treatment. The curves of electrical conductivity in wine showed a stable rising trend as a function of exposure time during ultrasound treatment. The longer the exposure time was, the larger the increasing in wine electrical conductivity. A sudden rise and a sudden drop of electrical conductivity were observed at the very beginning of ultrasound treatment and just when the ultrasound generator was turned off, respectively. Moreover, the sudden drop was followed by a stabilization of the electrical conductivity value at all the ultrasound powers, ultrasound frequencies and bath temperatures assayed. These stable values of electrical conductivity were markedly higher than those before ultrasound treatment being the effect exerted by temperature on wine electrical conductivity greater than those of the different ultrasound powers and frequencies assayed in our research (Zhang et al., 2016b). Furthermore, the four operational parameters studied had a significant effect on the changes of electrical conductivity during the ultrasound treatment, which suggested a positive correlation between them. Finally, we verified that the kinetic changes of electrical conductivity monitored in real time (Zhang et al., 2016b) were significantly different than the values measured at intervals (Zhang et al., 2016a) at the same exposure time and under the same operational conditions.

All in all, our results illustrate the potential of monitoring wine electrical conductivity as an important and simple indicator to control the status of chemical reactions in wine during ultrasound treatment in the future. However, considering the fact that the correlation between electrical conductivity and chemical reactions induced by ultrasound within wine remains to be demonstrated, we will conduct further research in this field.

OTHER APPLICATIONS OF ULTRASOUND IN THE WINEMAKING INDUSTRY

The main other application of ultrasound in the winemaking industry is as a simple and rapid extractive method of different wine compounds. In this regards, the extraction of aroma compounds from white wine using ultrasound along with a mixture of *n*-pentane and diethylether (1:2) as solvent has been reported (Hernanz et al., 1999). The advantages of using ultrasound over other extraction methods are higher reproducibility and possibility of simultaneous extraction of several compounds, which makes the ultrasound-assisted extraction an interesting alternative for the analysis of wine flavor components. Recently, ultrasound has also been reported to assist the extraction of different phenolic compounds present in grapes due to its simplicity and low cost. To be specific, ultrasound-assisted subcritical water extraction of proanthocyanidins from defatted grape seed markedly reduced the extraction time in comparison with other extraction techniques such as soxhlet's extraction and heat reflux extraction (Li et al., 2010). Tao et al. (2014c) developed several mathematical models to describe the ultrasound-assisted extraction of phenolic compounds from grape marc. Their results illustrated that exposure time was the sole parameter among the different assayed operational parameters (ultrasound energy density, ultrasound exposure time, and bath temperature) that affected the release of total phenolics. As aforementioned, these authors have also assayed the ultrasound-assisted extraction of phenolic compounds from oak chips (Tao et al., 2014b). The ultrasound-assisted subcritical water extraction also improved to 4.05% the proanthocyanidins yield under the optimized extraction conditions (temperature 145°C, extraction time 18 min, extraction pressure 14 MPa). Another extraction method using a sonication water bath was developed to recovery total phenolic compounds and anthocyanins from grape seed, being 29 min the optimal extraction time (Ghafoor et al., 2009). Furthermore, ultrasound-assisted extraction using ethanol-water as extracting solvent, allowed a quantitative and reproducible extraction of total phenolics, total anthocyanins and condensed tannins present in grape in a short time (6 min), compared with the classical phenolic compounds solvent extraction (60 min)

(Carrera et al., 2012). However, attention should be paid not only to the final concentration but also to the potential degradation of the ultrasonically extracted phenolic compounds. Zhang et al. (2015b) found a marked degradation of gallic acid during ultrasound-assisted extraction, which implies that the extraction yield should not be over-focused in actual extraction applications of ultrasound. Finally, ultrasound-assisted extraction of phenolic compounds from vineyard by-products has been assayed. In this sense, Delgado-Torre et al. (2012) obtained extracts with high phenolic content from vine-shots lignin by means of ultrasound-assisted extraction. These authors concluded that, from an oenological point of view, the concentration of representative phenols in extracts from vine-shoots (gallic acid, pyrogallol, catechin, acetovanillone, coniferaldehyde, sinapaldehyde) could be used to improve wine quality through oxidation/reduction reactions, similarly to the effect of oak chips and oak barrels in wine ageing.

Apart from the use as an extraction technique, ultrasound could have other applications in the wine industry related to the extreme cell disruption that ultrasound provokes in bacteria, yeast, fungi, algae and protozoa (García Martín and Sun, 2013). In this sense, ultrasound could be used to reduce the concentration of undesired microbes during both primary and secondary fermentation. Furthermore, ultrasound could be applied for cleaning and sanitizing of oak barrels and fermentation vats. These potential applications of ultrasound to wine microbiology have been largely reviewed and can be found elsewhere (Jiranek et al., 2008).

CONCLUSION

Ultrasound, a non-destructive technique, is nowadays applied in many industrial food processes. From an economical point of view, ultrasound equipment is not expensive compared to other technologies, so it could be implemented in the winemaking industry for wine ageing without the need to resort to a great initial investment. However, the actual potential of ultrasound to enhance wine quality remains to be demonstrated; results found in available literature to date are not very promising. In our research, we subjected a Cabernet Sauvignon red wine to ultrasound treatment. Our trials at different ultrasound powers (120, 150, 180, 210, 240, 270 and 300 W), frequencies (45, 80 and 100 kHz), exposure times (20, 40, 60, 80 and 100 min) and temperatures (20°C, 30°C, 40°C, 50°C and 60°C) significantly modified the concentration of total phenolic compounds while no significant effect on pH and total and volatile acidity was detected. Regarding wine color, the overall colorimetric difference in relation to the untreated wine reached 2.8 units, close to the color discrimination threshold for red wines (3 units). Nevertheless, further research is needed on a wider range of wine quality parameters since the current results do not allow assessing whether ultrasound positively modifies red wine quality.

Electron paramagnetic resonance spin trapping of free radicals with 5,5-dimethyl-1-pyrroline N-oxide allowed identifying the species of free radicals generated by ultrasound in red wine. Our results showed that the concentration of 1-hydroxylethyl free radical increased after ultrasound treatment, which demonstrated that ultrasound does trigger the generation of 1-hydroxylethyl free radicals into wine. Hydroxyl free radical was not detected in red wine because as soon it is generated, it reacts with ethanol in red wine to produce the novel 1-hydroxylethyl free radical.

The electrical conductivity of red wine was definitely influenced by ultrasound under all operational conditions employed, rising when increasing exposure time to ultrasound. The greatest increases on wine electrical conductivity were achieved when operating at the highest bath temperature. Therefore, the maximum values of electrical conductivities were reached during sonication at 60°C while the minimum values were found at 20°C. Furthermore, the kinetic changes of electrical conductivity monitored in real time were significantly different than the values measured at intervals at the same operational conditions. A sudden rise and a sudden drop of electrical conductivity were registered at the very beginning of ultrasound treatment and just when the ultrasound generator was turned off, respectively. Our results suggest that electrical conductivity could be used as an important and simple indicator to monitor the status of chemical reactions in wine during ultrasound treatment.

From our results and the information available in literature it could be concluded that the application of ultrasound to wine is not enough to accelerate the wine ageing process. We consider that it is worth to assay the combination of ultrasound with other ageing technologies already implemented in wineries, such as ageing in barrels (or oak chip addition), ageing on lees and microoxygenation. Since ageing on lees are generally performed together with ageing in barrels, both could be carried out simultaneously with ultrasound treatment. In this sense, if wood and lees compounds can fulfill their advantages during the wine ageing process, and ultrasound can accelerate the reactions that generally occur during natural ageing between these compounds and those within wine, the combination of these techniques with ultrasound could provide high-quality wines in a very short ageing time.

REFERENCES

Aadil R. M., Zeng X.A., Han Z., Sun D.W. (2013). Effects of ultrasound treatments on quality of grapefruit juice. *Food Chemistry,* 141, 3201–3206.

Bachmann J.A., Willkins R. (1937). Method of treatment for fermented and distilled beverages and the like. 2,086,891. USA Patent Application.

Bozalongo R., Carrillo J.D., Fernández Torroba M.A., Tena M.T. (2007). Analysis of French and American oak chips with different toasting degrees by headspace solid-phase microextraction-gas chromatography–mass spectrometry. *Journal of Chromatography A,* 1173, 10–17.

Buettner G.R. (1987). Spin Trapping: ESR parameters of spin adducts 1474 1528V. *Free Radical Biology and Medicine*, 3, 259–303.

Carrera C., Ruiz-Rodríguez A., Palma M., Barroso C.G. (2012). Ultrasound assisted extraction of phenolic compounds from grapes. *Analytica Chimica Acta,* 732, 100–104.

Castellanos M.M., Reyman D., Sieiro C., Calle P. (2001). ESR-spin trapping study on the sonochemistry of liquids in the presence of oxygen. Evidence for the superoxide radical anion formation. *Ultrasonics Sonochemistry,* 8, 17–22.

Chang A.C., Chen F.C. (2002). The application of 20 kHz ultrasonic waves to accelerate the aging of different wines. *Food Chemistry,* 79, 501–506.

Chang A.C. (2004). The effects of different accelerating techniques on maize wine maturation. *Food Chemistry,* 86, 61–68.

Chang A.C. (2005). Study of ultrasonic wave treatments for accelerating the aging process in a rice alcoholic beverage. *Food Chemistry,* 92, 337–342.

Colombié S., Latrille E., Sablayrolles J.M. (2007). On-line estimation of assimilable nitrogen by electrical conductivity measurement during alcoholic fermentation in enological conditions. *Journal of Bioscience and Bioengineering,* 103, 229–235.

Colombié S., Latrille E., Sablayrolles, J.M. (2008). Interest of on-line monitoring electrical conductivity during wine fermentation. *European Food Research and Technology,* 226, 1553–1557.

Cui Y., Lv W., Liu J., Wang B. (2012). Effect of different ending fermentation technologies on microbial-stability of Italian Riesling low alcohol sweet white wine. *Advanced Materials Research,* 393-395, 1165–1168.

Del Barrio-Galán, R., Pérez-Magariño, S., Ortega-Heras, M. (2011). Tecniques for improving or replacing ageing on lees of oak aged red wines: The effects on polysacharides and the phenolic composition. *Food Chemistry,* 127, 528–540.

Delgado-Torre M.P., Ferreiro-Vera C., Priego-Capote F., Pérez-Juan P.M., Luque De Castro M.D. (2012). Comparison of accelerated methods for the extraction of phenolic compounds from different vine-shoot cultivars. *Journal of Agricultural and Food Chemistry,* 60(12), 3051–3060.

Elias R.J, Andersen M.L, Skibsted L.H., Waterhouse A.L. (2009). Identification of free radical intermediates in oxidized wine using electron paramagnetic resonance spin trapping. *Journal of Agricultural and Food Chemistry,* 57, 4359–4365.

Fernández-Pachón M.S., Villaño D., García-Parrilla M.C., Troncoso A.M. (2004). Antioxidant activity of wines and relation with their polyphenolic composition. *Analytica Chimica Acta, 513*, 113–118.

García Martín J.F., Sun D.W. (2013). Ultrasound and electric fields as novel techniques for assisting the wine ageing process: The state-of-the-art research. Trends in Food Science and Technology, 33, 40–53.

García Martín J.F., Guillemet L., Feng C., Sun D.W. (2013). Cell viability and proteins release during ultrasound-assisted yeast lysis of light lees in model wine. *Food Chemistry*, 141, 934–939.

Ghafoor K., Choi Y.H., Jeon J.Y., Jo I.H. (2009). Optimization of ultrasound-assisted extraction of phenolic compounds, antioxidants, and anthocyanins from grape (Vitis vinifera) seeds. *Journal of Agricultural and Food Chemistry,* 57, 4988–4994.

Guerrero E.D., Mejías R.C., Marín R.N., Bejarano M.J.R., Dodero M.C.R., Barroso C.G. (2010). Accelerated aging of a Sherry wine vinegar on an industrial scale employing microoxygenation and oak chips. *European Food Research and Technology,* 232, 241-254.

Hernanz D., Heredia Mira F.J., Beltran R., Fernández M.A. (1999). Optimization of an extraction method of aroma compounds in white wine using ultrasound. *Talanta*, 50, 413–421.

Jackson R.S. (1994). Wine science: principles and applications. San Diego, USA, Academic Press.

Jambrak A.R., Mason T.J., Paniwnyk L., Lelas V. (2007). Ultrasonic effect on pH, electric conductivity, and tissue surface of button mushrooms,

Brussels sprouts and cauliflower. *Czech Journal of Food Sciences*, 25, 90–100.

Jiranek V., Grbin P., Yap A., Barnes M., Bates D. (2008). High power ultrasonics as a novel tool offering new opportunities for managing wine microbiology. *Biotechnology Letters,* 30, 1–6.

Kreitman G.Y., Cantu A., Waterhouse A.L., Elias R.J. (2013). Effect of metal chelators on the oxidative stability of model wine. *Journal of Agricultural and Food Chemistry,* 61, 9480–9487.

Kubo M., Sekiguchi K., Shibasaki-Kitakawa N., Yonemoto T. (2012). Kinetic model for formation of DMPO-OH in water under ultrasonic irradiation using EPR spin trapping method. *Research on Chemical Intermediates*, 38, 2191–2204.

Li C., Wang W., Yu H., Li J., Yang R. (2010). Ultrasound-assisted subcritical water extraction of proanthocyanidins from defatted grape seed and its antioxidant activity. *Zhongguo Zhong Yao Za Zhi,* 35(8), 967–972.

Liu L, Loira I., Morata A., Suárez-Lepe J.A., González M. C., Rauhut D. (2016). Shortening the ageing on lees process in wines by using ultrasound and microwave treatments both combined with stirring and abrasion techniques. *European Food Research and Technology, in press* (DOI:10.1007/s00217-015-2566-z).

Lucena A.P.S., Nascimento R.J.B., Maciel J.A.C., Tavares J.X., Barbosa-Filho J.M., Oliveira E.J. (2010). Antioxidant activity and phenolics content of selected Brazilian wines. *Journal of Food Composition and Analysis,* 23, 30–36.

Masuzawa N., Ohdaira E., Massao I. (2000). Effects of ultrasonic irradiation on phenolic compounds in wine. *Japanese Journal of Applied Physics*, 39, 2978–2979.

Makino K., Mossoba M.M., Riesz P. (1983). Chemical effects of ultrasound on aqueous solutions. Formation of hydroxyl radicals and hydrogen atoms. *The Journal of Physical Chemistry,* 87, 1369–1377.

Morales M. L., Benitez B., Troncoso A. M. (2004). Accelerated aging of wine vinegars with oak chips: evaluation of wood flavour compounds. *Food Chemistry*, 88, 305–315.

Robinson J. (2006). The Oxford companion to wine. *Oxford University Press,* Oxford, UK.

Sánchez-Moreno C., Larrauri J.A., Saura-Calixto F. (1999). Free radical scavenging capacity of selected red, rose and white wines. *Journal of the Science of Food and Agriculture*, 79, 1301–1304.

Singleton V.L., Draper D.E. (1963). Ultrasonic treatment with gas purging as a quick aging treatment for wine. *American Journal of Enology and Viticulture*, 14, 23–35.

Tao Y., García J.F., Sun D.W. (2014a). Advances in wine aging technologies for Enhancing wine quality and accelerating wine aging process. *Critical Reviews in Food Science and Nutrition,* 54, 817–835.

Tao Y., Zhang Z., Sun D.W. (2014b). Experimental and modeling studies of ultrasound-assisted release of phenolics from oak chips into model wine. *Ultrasonics Sonochemistry*, 21, 1839–1848.

Tao Y., Zhang Z., Sun D.W. (2014c). Kinetic modeling of ultrasound-assisted extraction of phenolic compounds from grape marc: Influence of acoustic energy density and temperature. *Ultrasonics Sonochemistry*, 21, 1461–1469.

Zamora F. (2002). La crianza del vino tinto con lías, una nueva tendencia. *Enólogos,* 19, 26–25.

Zhang Q.A., Shen Y., Fan X.H., García Martín J.F., Wang X., Song Y. (2015a). Free radical generation induced by ultrasound in red wine and model wine: An EPR spin-trapping study. *Ultrasonics Sonochemistry*, 27, 96–101.

Zhang Q.A., Shen H., Fan X.H., Shen Y., Wang X., Song Y. (2015b). Changes of gallic acid mediated by ultrasound in a model extraction solution. *Ultrasonics Sonochemistry*, 22,149–154.

Zhang Q.A., Shen Y., Fan X.H., García Martín J.F. (2016a). Preliminary study of the effect of ultrasound on physicochemical properties of red wine. *CyTA Journal of Food*, 14(1), 55–64.

Zhang Q.A., Shen Y., Fan X.H., Yana Y.Y., García Martín J.F. (2016b). Online monitoring of electrical conductivity of wine induced by ultrasound. *CyTA Journal of Food*, 14(3), 496–501.

Zoecklein B. K., Fugelsang B., Gump B., Nury, F. (1995). Wine analysis and production. New York, NY: Chapman and Hall.

In: Applications of Ultrasound in the Beverage… ISBN: 978-1-63485-069-8
Editor: Juan Francisco García Martín

Chapter 6

ULTRASOUND FOR ACCELERATING THE WINE AGEING PROCESS: A WINETASTER'S VIEW

Dermot Nolan*
Master of Wine

A few years ago I was asked to take part as an objective wine taster in a major "Research for SMEs" project funded by the European Commission's Seventh Framework Programme (FP7) which had the backing of six companies, ranging from wine producers to equipment manufacturers from across Europe. The idea behind the project was to determine if it was possible to age a wine using ultrasound. The methods of this work are described elsewhere so here I am going to outline my thoughts in relation to the ageing of wine and whether ultrasound worked.

In the world of wine terms are used rather loosely and, at times, confusingly. A wine can be aged or matured and these terms refer to different stages of development. Maturation occurs after fermentation and is used to mean time spent in tanks or barrels. During this stage of development a number of changes can occur.

Ageing is usually understood to mean time spent in bottle and in this type of development a wine can proceed from youth, through to maturity and, eventually, to a fading away of any worthwhile flavours.

* Corresponding Author Email: dermot@dermotnolan.ie.

Confusingly, wine tasters can refer to a wine which has gone through extensive ageing as being mature yet also call a wine which has undergone maturation in the winery as youthful!

In general, maturation in the winery prepares a wine for its useful lifetime, the time in which the wine is good to drink. For the vast majority of wines this period is about 2 years at most. The maturation process stabilises the wine and should enable the wine to be drunk pleasantly and enjoyably on its commercial release. However, for some wines this maturation can last for over 100 years!

Bottle ageing allows a wine's flavours to develop in more complex ways, adding layers of unexpected flavours which add to the enjoyment of the wine. This ageing can take place over a number of years with the development of flavours taking place over some 20 years or more. There then comes a point in time when the wine is deemed to be drinking at its best but this period can last for another 20, or 50 or even 100 years.

For a wine to be so age worthy it must have a number of preservatives and be made in such a way as to be able to develop new flavours. In white wines, the wine must have a low pH or high levels of residual sugar to help protect it against all forms of microbial spoilage. Red wines usually have tannins as their main preservative, with pH also being important. In all cases sulfite levels are important as both anti-oxidants and anti-microbials.

Not all wine grapes are capable of making wines which will age. In some instances this is because the grape cannot offer sufficient acidity or tannin to protect the wine but in the majority of cases it is because the general belief is that wines from such grapes do not develop pleasing flavours with time.

Wine tasters distinguish between three levels of flavour: primary, secondary and tertiary. Primary flavours are related directly to the grape such as blackcurrant in wines made from cabernet sauvignon. Secondary flavours arise from winery procedures such as the vanillin extracted from new oak barrels, or sulfite characters which arise from reductive maturation. Some secondary flavours are considered detrimental such as high volatile acidity or the rank flavour of wine which has been affected by the yeast *Brettanomyces*. Tertiary flavours are those which arise during bottle ageing. Red wines typically develop brown fruit flavours such as figs, prunes and so forth, as well as tobacco leaf, chocolate and other characters. White wines tend towards waxy, lanolin type flavours but also honey, toast and so forth.

The greater the propensity for a wine to develop all three flavours then the higher the esteem in which it is held. Such wines are complex and highly sought after and can demand exceptionally high prices.

Now, many of the preservatives in wine can act as a barrier to enjoyable drinking. Tannins, for example, react with the proteins present in saliva to produce a mouth-drying effect and this can make a wine seem astringent. With time, tannins eventually reduce as polymerisation in the bottle results in molecular weights which are unstable in wine which then causes the tannin complexes to precipitate from solution, forming the sediment in the bottle. As these complexes also contain anthocyans the wine loses colour over time.

With white wines the situation is somewhat different. Acids can make a wine fresh but can also make a wine sour or tart. Sugar can soften the taste of the acidity but can also make a wine heavy or cloying. Neither sugar nor acid levels change that much with time so the beneficial ageing of white wines requires that positive flavours evolve over time in the bottle.

So, ultimately there are two questions to ask about what might be termed artificial ageing of wine, by any means including ultrasound. Can artificial ageing make a slightly tough young wine drinkable and can artificial ageing enable a more rapid evolution of positive flavours in the wine?

It is worth noting that in the case of the first question one has to ask why the wine was not better made. If a wine is intended for fairly immediate consumption then why are there any undesirable characteristics there in the first place? Too much tannin in a young red wine can be avoided by better winemaking techniques which are already available to winemakers without the expense of an artificial ageing machine.

In the case of wines made for ageing (vins de gardes as the French would have it) one has to ask do we even know now what reactions take place to bring about the desired flavour development in the wine. This is an area which, as I understand it, is little understood today.

So, these questions were always in mind as I tasted wines which had undergone various ultrasound treatments. In general I found no positive changes to any of the wines. In the vast majority of cases there was no detrimental change either so my conclusion is that ultrasound treatments made no real difference to the flavour profile of a wine.

Now in some cases the wines seemed slightly less enjoyable than untreated control samples but this could be due to changes in such aspects of total acidity or pH or perhaps due to increased polymerisation of tannins. However, I was not privy to analytical data which might have pointed in this direction. On a number of occasions some wines developed a slight "burnt" flavour, or a slight metallic character but not frequently enough to be significant. However, it is worth pointing out that any process which causes such changes is to be avoided.

My overall feeling was that while the exercise was interesting it had no measurable effect and, further, I am not sure that there is a good existential reason for such treatments. Consumers who like aged wines are well prepared to cellar these wines themselves, to assume the cost which this entails and to wait for the wine to reach a point in its development where they deem the wine to be at its best. Such wines make up a tiny, tiny percentage of all wines produced and the requirement for accelerated ageing does not, in my opinion, exist in this niche market segment.

The vast majority of wines will keep (that is stay in bottle without developing aged flavours) for a good 2 to 3 years but their consumers typically consume them within 48 hours of purchase and about 80% of all production is consumed within 18 months of commercial release. This suggests that these wines are already mature enough to drink and that the flavours they offer are attractive to consumers, so why try to change? Any form of artificial ageing is likely to be an expensive process and would probably add considerably to the cost of wine. Such added cost without any tangible and substantial improvement is not likely to be viewed favourably by producers or consumers alike.

In a sense, if it ain't broke, don't fix it. I feel, overall, that the research was trying to answer a question which does not actually exist.

In: Applications of Ultrasound in the Beverage… ISBN: 978-1-63485-069-8
Editor: Juan Francisco García Martín

Chapter 7

ULTRASOUND APPLICATION IN WINEMAKING FOR A LOW IMPACT ENOLOGY

Paola Ferraretto and Emilio Celotti*
University of Udine, Department of Food Sciences, Udine, Italy

ABSTRACT

The ultrasound technology, when applied to a liquid medium, causes a physical effect known as cavitation, with chemical and mechanical consequences. The effects of this technology are interesting in food industry; therefore we investigated new potential applications for the vinification process. We studied the phenolic compounds extracted from grape based on the disruption of cell wall, due to pressure alternance and cavitation. Tests carried out for few minutes of treatment at 20 kHz on different grapes, showed an improvement in the extraction of polyphenolic substances with a reduction in the time of classic maceration. Moreover, the impact of ultrasonic waves on tannins and anthocyanins, which play a role on color, stability and organoleptic properties of wine were evaluated on different red wines. It was found that a few minutes of ultrasonic treatment promote an increase of catechins, positive effects on color stability and astringency.

* Corresponding Author Email: emilio.celotti@uniud.it.

Keywords: ultrasound, winemaking, phenolic compounds, extraction, aging, color stability, sustainability

INTRODUCTION

Ultrasound represents an innovative technology for the food industry. This technology has become an alternative food processing technology to produce high quality and safe products, as well as to improve energy and processing efficiency.

The ultrasound technology is based on mechanical waves at a frequency higher than the upper limit of human hearing (>16 kHz). Being sound waves, ultrasound is transmitted through any substance, solid, liquid or gas, which possesses elastic properties. These waves travel either through the bulk of a material or on its surface at a speed which is characteristic of the nature of the wave and the material through which it is propagating (Povey and McClements, 1988; Mason, 1998; Knorr et al., 2004). In the food industry, ultrasound can be classified into two frequency ranges: high frequency ultrasound (100 kHz – 1 MHz) and power ultrasound (16 – 100 kHz).

High frequency ultrasound is commonly used as an analytical technique to provide information on the physicochemical properties of food such as ripeness, sugar content, acidity, etc. (Demirdöven and Baysal, 2009). The application of power ultrasound is so large that it is used to modify either physical or chemical properties of food (McClements, 1995). Power ultrasound has been used for many years to generate emulsions, disrupt cells, and disperse aggregated materials; more recently, some applications have been identified with greater potential for future development, i.e., modification and control of crystallization processes, degassing of liquid foods, enzymes inactivation, enhanced drying and filtration and the induction of oxidation reactions (Knorr et al., 2004; McClements, 1995; Roberts, 1993; Zheng and Sun, 2006).

The fundamental effect of ultrasound on a continuum fluid is to impose acoustic pressure in addition to the hydrostatic pressure already acting on the medium. The combination of factors such as heat, pressure and turbulence, is used to accelerate mass transfer in chemical reactions, creates new reaction pathways, break down and dislodge particles or even generate different products from those obtained under conventional conditions (Suslick, 1998).

The use of ultrasound in an industrial process requires two main factors: a liquid medium and a source of high-energy vibrations. Ultrasound is propagated via a series of compression and rarefaction waves induced in the molecules of the medium through which it passes. At sufficiently high power, the rarefaction cycle may exceed the attractive forces of the liquid molecules, and cavitation bubbles will form. Such bubbles grow rapidly, followed by a violent collapse when they reach a critical size. This process is known as transient cavitation.

In aqueous systems, at the ultrasonic frequency of 20 kHz, each cavitation bubble collapse can locally generate extreme temperature (up of 5000 K) and pressure (up of 1000 atmospheres) (Gong and Hart, 1998). At a solid and liquid interface, the water jet formed by an imploding transient cavitation is also a factor that may contribute to some changes in the overall properties of a food product. It is noteworthy that all the chemical and physical effects of ultrasounds are microscopic. However, the interplay of these cavitation-induced chemical and physical activities with food is manifested through macroscopic changes that can be perceived by consumers and may be expressed through terms such as texture, color and flavor (Feng and Lee, 2011).

The properties of power ultrasound waves have generated interest in the food industry, since the promotion of physical and chemical reactions can lead to a strategic advantage in processing. Since the '90s, ultrasound treatments have become a widely applicable alternative in the processing industry (Bates and Patist, 2010); ultrasound is applied to food technology for its mechanical and/or chemical effects on homogenization, mixing, extraction, filtration, crystallization, dehydration, fermentation, degassing, defoaming, particle size reduction, temporary or permanent changes of viscosity, modulation of growth of living cells, cells disruption and aggregate materials dispersion, inactivation of microorganisms and enzymes, sterilization of equipment (Mason et al., 1996; Chemat et al., 2011; Chandrapala et al., 2012; Tao and Sun, 2015). For these reasons, the effects of this technology are interesting in the food industry, but there has not been much research done on fermented beverages, especially regarding the winemaking process.

The potential applications of high power ultrasound in the wine industry have been summarized by Yap et al. (2007) and Jiranek et al. (2008). They include the management of microorganisms at different stages of wine production, color and flavor extraction, cold stabilization, protein stabilization, enhancement of yeast cell autolysis, defoaming, degassing, and barrel and tank sanitization.

Guerrero et al. (2001) have published a study on the ultrasound effects on *Saccharomyces cerevisiae* survival after application of wave amplitude, in the range of 71 to 110 μm and a frequency of 20 kHz. The microscopic observations determined that ultrasound caused puncturing of cell walls with leakage of content as well as damage at subcellular level.

Ultrasound was also applied with the aim of eliminating solid residues and *Brettanomyces* contamination of oak barrels, and it proved efficacy in barrel disinfection (Yap et al., 2008). Cavitus has launched its patented ultrasonic barrel cleaning and disinfection system into the wine industry (Clark, 2008).

Ultrasound has recently been applied to evaluate the possible effect on color and flavor extraction at different stages of the winemaking process (Bates and Patist, 2010; Lieu and Le, 2010; El Darra et al., 2013; Ferraretto et al., 2013): red color and aromatic compounds are localized in the skin cells and their release is facilitated by mechanical actions, disruption of tissues and cells, temperature and alcohol. The ultrasound cavitation, causing disruption of the cell walls and membranes, would be able to facilitate the release of these compounds. Application of ultrasound is considered an alternative approach to traditional methods.

In addition to this, in the last few years, some researchers have focused their attention on wine aging accelerated by the application of physical methods, unlike chemicals, such as using ultrasound (Chang and Chen, 2002; Chang, 2005; Masuzawa et al., 2000; García Martín and Sun, 2013; Zhang et al., 2016; Cacciola et al., 2013; García Martín et al., 2013).

Red grapes and their phenolic compounds are the most important raw material for winemaking. Anthocyanins and tannins are the main phenolic compounds responsible for color, taste, and aging properties of the red wine. Both compounds are found in berry skins, in the cell vacuoles, although tannins are also found in the seeds. The extraction of skin phenolic compounds during fermentative maceration is essentially a diffusion process, and the rate of extraction is influenced by the skin phenolic concentration, the composition of berry cell walls, and the technological process applied to the winery (Busse-Valverde et al., 2011).

Several studies have focused on techniques able to weaken the cell walls and facilitate the diffusion of phenolic compounds during vinification (Sun et al., 2001; Sacchi et al., 2005). Thermovinification enhances the extraction of phenolic compounds: this technique permits a solubilization and diffusivity of cell components, but it may cause degradation of the phenolic substance. The thermal treatment increases volatilization of aromatic compounds, reducing the quality of the wine (El Darra et al., 2013).

Innovative technologies, including ultrasound, have been recently tested to improve the extraction of phenolic compounds from grape berries, grape mash, grape skin, and seeds (Lieu and Le, 2010; Chemat et al., 2011; El Darra et al., 2013; Ferraretto et al., 2013). Ultrasound can effectively enhance the extraction of intracellular compound from different plants. El Darra et al. (2013) summarizes what literature has reported: the ultrasound application at 20-35 kHz enhances the extraction of polyphenols from red grape residues and from grape seeds of phenolic and other bioactive compounds from grape must and the flavonoid extraction from grape skin without the undesirable degradation of phenolics.

The extraction improvement is caused by the break of cell membranes: under ultrasonic irradiation, the mass transfer rate can be enhanced markedly. The microstreaming caused by ultrasound has a mechanical effect on the surface of raw materials, which can destroy the surface and reduce the particle size. Moreover, the high temperature and pressure can also facilitate the destruction of material surface. Therefore, the area exposed to the ultrasonic field increases and the compounds inside the raw materials tend to release into solvent (Tao and Sun, 2015).

Owing to their essential contribution to wine sensory properties (color, flavor, astringency, and bitterness) and to their role during aging (Spranger et al., 2004) the phenolic compounds are extremely important in enology.

The parameters for the evaluation of red wines coming from red grape varieties, considering the different methods of color extraction (i.e., maceration, thermovinification, carbonic maceration, pectinolytic enzymes), are strictly related to the quality and the quantity of the phenolic compounds in the raw material, but the main constraint of the technology used is the polymerization kinetic that influences the times of wine aging.

Ultrasound could also promote all the reactions which occur during the aging process, such as oxidative reactions (involved or not molecular oxygen) and reductive reactions.

One of the most important changes during aging is a progressive increase and stabilization of the color due to copigment anthocyanin complexes, the formation of new pigments and the progressive formation of both tannin-tannin and anthocyanin-tannin complexes (Jackson, 2008; Boulton, 2001).

Several studies have literally illustrated the potential of alcoholic beverages aging by chemical methods (Boulton, 2001; Castellari et al., 2000), while very few experiments have been done applying physical methods, such as ultrasound (Chang, 2005). According to Chang and Chen (2002) the application of ultrasonic waves with a frequency of 20 kHz facilitates the aging of rice wine more than the conventional one, without affecting the final quality of the product. On the contrary the application of ultrasound treatment on maize wine promotes the aging, but the final quality is not comparable with the traditional aging quality (Chang, 2004).

Masuzawa et al. (2000) described the ultrasound effects on phenolic compounds of red wine. From the results obtained, the authors confirm an effect of polymerization of polyphenolic compounds in red wine promoted by ultrasound at low sound pressure levels. Ultrasound effects on red wine physicochemical properties have been recently investigated (Zhang et al., 2016). The authors have reported that different conditions of ultrasonic treatment have significantly changed the concentration of total phenolic compounds and the electrical conductivity. Therefore, this physical technology, modifying the characteristics of red wine, could be a promising method for accelerating the wine aging process (García Martín and Sun, 2013).

Ultrasound power provides high temperature and high pressure leading to the modification of chemical reactions (Suslick, 1989), such as fragmentation and subsequent recombination of polymers (Chang, 2005). The increase of tannins and anthocyanins concentration in wine, the main phenolic compounds responsible for flavor, color and aging, leads to wines with a higher aging potential. Therefore, the application of a polyphenol extraction technique before fermentation could enhance the further aging process (García Martín and Sun, 2013).

Ultrasound could also be used to accelerate the release of protective colloids from lees and stabilize wines in a very short time (Cacciola et al., 2013; García Martín et al., 2013). The ultrasound cavitation would be able to facilitate the release of these compounds, causing disruption of cell walls and membranes.

Ferraretto and Celotti (2016) have recently explored the stabilizing effects of ultrasound on tannin-anthocyanins polymers, in order to understand how to speed up the aging reactions and hence reduce the time between production and consumption.

Ultrasound, as a relatively low-cost, non-hazardous and environmental friendly technology, is commonly used in the food industry (Mason et al., 1996; Sun and Li, 2003; Zheng and Sun, 2006), hence its possible application in the wine industry might become an important technological innovation by speeding up some slow reactions required in the winemaking process.

We think that a possible application of ultrasound in the wine industry could be an important technological innovation that would fasten the thermodynamically possible reactions, which are required in the winemaking process, although they are kinetically slow. In this chapter we attempt to present our ultrasound applications to grapes as raw material and to the vinification process. At first, we studied the extraction of phenolic compounds from grapes obtained by the disruption of the cell wall due to pressure alternance and cavitation and then we evaluated the application on red wine in order to accelerate the process of aging.

METHODS

Ultrasonic Equipment

An ultrasonic processor (SONOPLUS HD 2200, Bandelin electronic, Berlin, Germany) with 13 mm sonotrode probe, made of titanium, was used for sonication.

Samples were processed in a continuous sonication at a constant frequency of 20 kHz. The energy input was controlled by setting the amplitude of the sonicator probe; the total nominal output was 200 W. The ultrasound probe was submerged to a depth of 50 or 20 mm in the sample on the basis of the experiments.

Its basic components are the power generator and the transducer that converts electrical power into mechanical vibrations. A schematic model is shown in Figure 1 and manufacturer details under it.

- Generator of pulse electric field (GM2200): Transforms low-frequency voltage of 50 Hz into high-frequency voltage of 20 kHz. It contains numerical display showing amplitude (%), pulsation mode (continuous or timed) and time (maximum 99 min).
- Transducer (UW2200): Transforms electrical energy delivered from the generator into mechanical vibrations of 20 kHz.
- Probe (SH 213G): Probe increases the amplitude by its specially designed shape.
- Replaceable tip (TT13): 13 mm diameter. Made of titanium alloy (Ti-Al6-V4) transmit mechanical longitudinal waves into the sample.
- Vessels used to process the samples were Pyrex glass 250mL beaker, making sure that the probe was centered.

- Amplitude used in the different treatments is equivalent to: 30% (51 μm), 60% (102 μm) and 90% (153μm). The higher the amplitude, the more intense sonication was developed.

The amplitude levels and the processing times were varied on the basis of the experiments and on the tests carried out prior to this study. The final treatment temperature was measured for all samples.

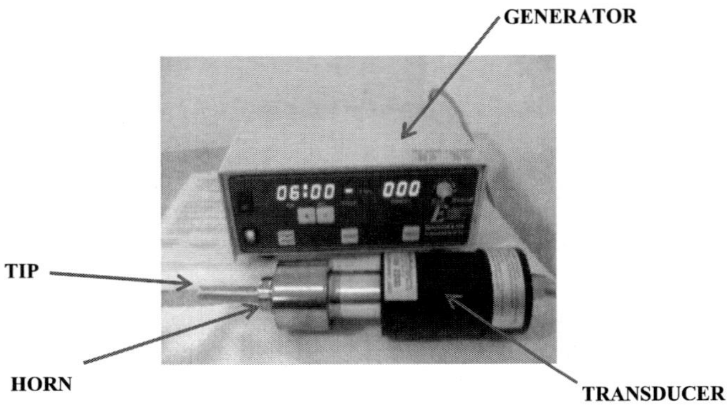

Figure 1. Ultrasonic equipment - basic components for SONOPLUS HD 2200.

Part 1: Preliminary Study of the Effects of Ultrasound on Red Grape Maceration

Phenolics Extraction from the Red Grape Berry Skin

Samples: a wide set (nr. 40) of different varieties of red grapes sampled at technological maturity was evaluated for the tests conducted in laboratory. They were sampled directly upon delivery in the cellar, in order to have a representative and real sampling in relation to the ripening degree of the grapes delivered to the cellar.

A representative sample (250 g), a stalk removed and manually crushed, was subjected to the treatment in a beaker of 250 mL. To obtain a broad statistical basis, it was decided to increase the number of varieties in a single test to assess the extent of the extraction. The polyphenols extracted were evaluated in the liquid obtained by centrifugation from the treated grapes (Figure 2).

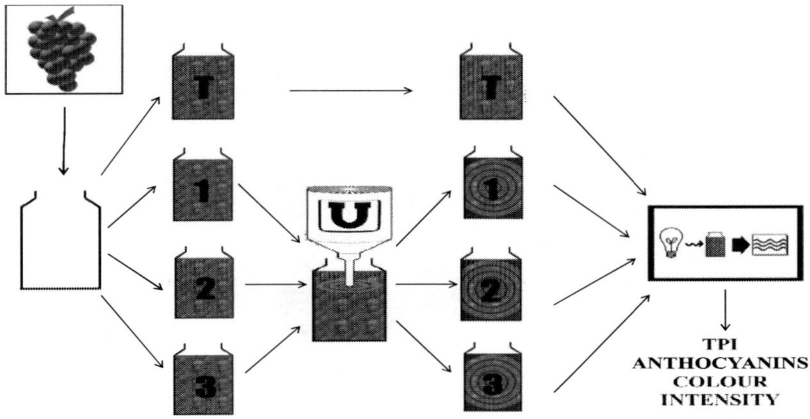

Figure 2. Plan of the ultrasound treatment trial on crushed grapes of different varieties of red grapes harvested at technological maturity and compared with the corresponding untreated sample (T). TPI: Total Polyphenol Index.

Treatment conditions: the conditions promising the best results in terms of extraction following a preliminary test (treatment time from 1 to 5 minutes at three different levels of amplitude – 30, 60 and 90%) were chosen for the laboratory tests. The treatment conditions selected to study phenolics extraction from red grape were 3 and 5 minutes at the amplitude of 90%. At 5 minutes of treatment time, the maximum sample temperature reached was 47.6°C for the amplitude level of 90%.

To evaluate the immediate effect on the color, the treatment conditions were 2 minutes at different amplitudes (20, 40, 60 and 80%).

Micro-Vinification Trials

Samples: 800g of crushed samples of the same grapevine variety (Merlot), but with a different polyphenolic content were used to perform maceration and fermentation trial in the cellar. The crushed grapes were previously subjected to ultrasonic treatment and then micro-vinified and racked sequentially. Experiments were performed in duplicate.

A reference sample, not treated with ultrasound, was analyzed for all the tests in order to have the sample that simulated the traditional maceration. All the samples, including the reference, were placed in open containers (2 L) and placed in the same conditions of maceration-fermentation in order to make comparable the phenolics extraction.

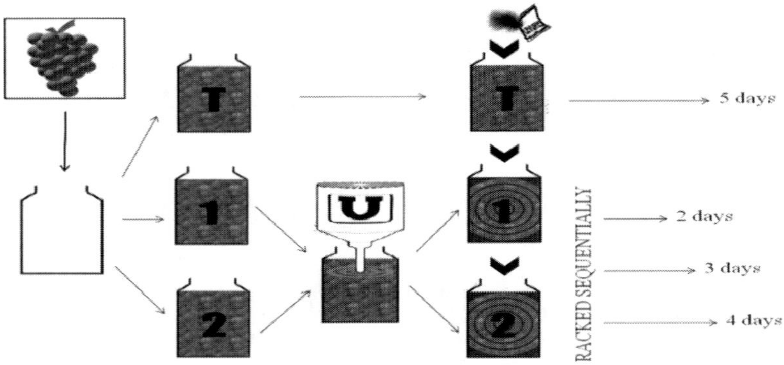

Figure 3. Plan of the comparison trials between classic maceration (T) and crushed red grapes subjected to ultrasound treatment, microvinified and racked sequentially.

The maceration conditions provided were: a constant temperature of 25°C, three manual pumping over and cap punching a day, 60 ppm SO_2, inoculum of dry yeast for oenological use *Saccharomyces Cerevisiae* on the first day and at the end of fermentation the separation of the solid parts was carried out with a manual press.

The effect on wine color was studied on micro-vinification trials of the samples performed directly in a cellar. The maceration time for the ultrasound treated samples was 2, 3 and 4 days, while for the control sample (untreated) it was 5 days (Figure 3).

Ultrasound treatment conditions: a treatment of 5 minutes at the amplitude level of 90% was carried out to assess ultrasound effects on the maceration kinetics. At 5 minutes of treatment time, the maximum sample temperature reached was 46.9°C.

Analytical Determinations

All the analytical determinations were conducted on grape samples right after the crushing (reference value) and right after the treatment (test results) once centrifuged at 3000 rpm for 10 minutes. All the analysis was performed in triplicate.

Determination of Total Polyphenol Index (TPI)

TPI is determined using the easiest method for a quickly estimation proposed by Ribéreau-Gayon (1970) by spectrophotometry at 280 nm.

Samples are placed in the spectrometer (with an appropriate dilution with distilled water), a light is passed through the liquid at a wavelength of 280 nm (using a 10 mm optical path), and the light is collected and compared to the original light source.

This value is based on the characteristic absorption of the benzene cycles of the majority of phenols at 280 nm. This test presents a number of advantages, including speed and reproducibility. Absorbance is also called optical density (OD).

$$I280 = OD \times dilution$$

Determination of Color Intensity (CI) and Hue (T)

Color intensity and hue are determined using the method proposed by Glories (1984) and Sudraud (1958). Red wines have a maximum at 520 nm, due to anthocyanins and their flavylium combinations, and a minimum in the region of 420 nm. A rapid approach to color analysis in winemaking requires optical density measurements at 420 and 520 nm, with an additional measurement at 620 nm to include the blue component in young red wines.

The spectrophotometric measurements must be made on a 1 mm optical path, using undiluted wine. These measurements are used to calculate the values used to describe wine color (Glories, 1984). Color intensity represents the amount of color.

$$CI = OD\ 420 + OD\ 520 + OD\ 620$$

The hue indicates the development of a color towards orange. The value increases throughout aging.

$$T = OD\ 420/OD\ 520$$

Determination of Anthocyanins Content

The anthocyanins content are measured according to Ribéreau-Gayon and Stonestreet (1965). The polymeric pigments are analyzed after the bisulphite bleaching of anthocyanin monomers carried out on an acidified sample ($pH < 1$).

The SO_2 bleaching procedure requires the preparation of a mixture containing 1 ml of sample or wine, 1 ml of EtOH 0.1% HCl and 20 ml of HCl at 2%. Then, in 2.5 mL of the mixture is added 1 ml of H_2O and in other 2.5 mL of the mixture is added 1 mL of potassium bisulfite solution (20%). The absorbance of each solution is measured, after 15 minutes, at 520 nm on a 10 mm optical path using distilled water as the control. By comparison with a standardized anthocyanin solution, the concentration is:

C (mg/L) = (OD potassium bi sulphate – OD distilled water) × 875 × dilution

To be able to compare the increase in compounds extracted versus the classical vinification, a reference sample followed extraction with a wine-like solution (5 g/L tartaric acid, 12% (v/v) ethanol, pH was adjusted to 3.2 with NaOH) or methanol (1% acidified with chloridric acid) for 4 hours, then centrifuged and analyzed.

Anthocyanins content and total polyphenols index - Abs 280 nm (of micro-vinified and reference samples were analyzed daily.

Determination of Condensed Tannins

The proanthocyanidin concentration was determined using the methodology proposed by Bate-Smith (1954). The principle is based on proanthocyanidin depolymerization through the breakdown of their intra-flavonols bonds in an acidic heat medium.

Each wine sample was diluted at a ratio of 1/50 (v/v) with deionized water. In two separate test tubes, 2 mL of the diluted sample and 6 mL of hydrochloric hydrochloric acid–butanol were added. One test tube (reaction tube) was placed in a water bath at 100°C for 30 min, and the other test tube (blank tube) was left to stand in the dark for the same time. After 30 min, when the heated reaction tube was cooled, the absorbance of each test tube was measured in a spectrophotometer at 550 nm using deionized water as the blank. The absorbance difference was multiplied by the factor 0.1736 and by dilutions (50), and the concentration of proanthocyanidins was expressed in g/L.

Determination of Polymerized Pigments Index

The polymerized pigments index was determined using the method proposed by Glories (1978). This index represents the contribution to the red color of condensed tannins and polymerized forms of anthocyanins insensitive to bleaching by sulfur dioxide. To 5 mL of sample shall be added, in a beaker, 45 mL of tartaric buffer pH 3.2 (5 g/L of tartaric acid brought to pH 3.2 with 4N NaOH) and 0.2 mL of 20% solution of potassium metabisulfite; in a second beaker, 5 mL sample and 45 mL of tartaric buffer are added with 0.2 mL of water. After 5 minutes the optical density is read, using deionized water as blank, at 420 nm and 520 nm for both tests.

$$\text{Polymerized pigments index (PPI)} = \frac{DO\ 420_{(SO_2)} + DO\ 520_{(SO_2)}}{DO\ 420_{(H_2O)} + DO\ 520_{(H_2O)}}$$

Anthocyanins content and total polyphenols index of micro-vinified and reference samples were analyzed daily. In addition to these analyses, tannins content and polymerized pigments index were analyzed after racking (after 2, 3 and 4 days for the treated sample and 5 days for the control).

Part 2: Preliminary Study of the Effects of Ultrasound on Red Wine Polyphenols

Treatment Conditions

Experimental tests, carried out to assess the possible effects of ultrasound on red wines, have been set according to different work plans.

For preliminary tests (samples 1 and 2), all treatment combinations (time and amplitude %) were analyzed with the response surface method (RSM: Response Surface Methodology) in a model of experimental design compound central face-centered (CCF: Central Composite Face-Centered) in order to optimize the number of experiments and operating conditions, and to obtain the greatest number of data with the minimum number of samples.

The experimental design for the tests was established with the statistic software Modde 8.0.2 of Umetrics. The experimental design provides the randomized treatment of 11 samples with three replicates of the central point (3 minutes at 60% of amplitude). The level of amplitude varied between 30% and 90% and the time of treatment between 1 and 5 minutes; the reference sample, not subject to treatment, was evaluated separately. The conditions for sample 3 were 1, 3 and 5 min at 30%, 60% and 90% (51, 102, 153 μm) of amplitude and each treatment was replicated 3 times.

Samples

The tests have been performed on young red wines (4 months after racking) immediately after the treatment: 200 mL of wine for every sample have been treated. The wines used were the following:

Sample 1: red table wine from Veneto region;
Sample 2: red table wine, Cabernet Sauvignon, from Friuli Venezia Giulia region;

Sample 3: red DOC (Protected Designation of Origin) wine, Raboso, from Veneto region.

Analytical Methods

Effects of ultrasound treatment were evaluated using the Glories' index (Glories, 1978) which show the status of phenolic compounds in red wine, mainly based on the absorbance measured using a spectrophotometer.

All the treated samples, as the reference ones, have been evaluated after the treatment for: total phenolics index - Abs 280 nm, color intensity, and anthocyanin content described in Part 1. In addition, hydrochloric acid index, ethanol index and gelatin index and catechins content (Zironi et al., 1992) were analyzed.

Determination of Hydrochloric Acid Index

The hydrochloric acid index was determined using the method proposed by Glories (1978). The measuring principle of the HCl index is based on the fact that procyanidins become unstable in a highly acidic environment. The speed of condensation depends on the degree of polymerization.

The procedure required to add to 10 mL of wine, 15 mL of 12 N HCl and 5 mL of water. This sample solution is diluted to thirty times its original volume and the optical density (d_0) at 280 nm is measured immediately in a 10 mm optical path.

The same measurement is then repeated on a sample where the acidic solution was left standing for 7 and 24 hours before dilution and it was then spinned. Two new values (d_1 and d_2) were obtained.

The HCl index is given by the equations:

$$\mathbf{I_7}(\mathrm{HCl}) = (d_0 - d_1)/d_0 \times 100$$
$$\mathbf{I_{24}}(\mathrm{HCl}) = (d_0 - d_2)/d_0 \times 100$$

The HCl index, reflects the state of polymerization of tannins in the wine, which, in turn, depends on the aging conditions.

Determination of Ethanol Index

The ethanol index was determined using the method proposed by Glories (1978).

The Ethanol index estimates the degree of combination of the tannins with the polysaccharides and their colloidal state. This index precipitates condensed anthocyanin polysaccharides by the addition of an excess ethanol.

To determine the index, 1 mL of sample is added, in a beaker, to 9 mL of 96% ethanol and 90 mL of distilled water and the OD 280 nm against water (d_0) is measured. In a second beaker, 5 mL of sample are added to 45 mL of 96% Ethanol. After 24 hours samples are subjected to centrifugation (3000g x 10') and after diluting the supernatant 10 times, the optical density at 280 nm against water (d_{24}) is measured.

$$\text{Ethanol index} = \frac{d_0 - d_{24}}{d_0} \times 100$$

The Ethanol index depends on the age of the wine and is related to the organoleptic characteristics such as softness and fullness of the wine. The value rises with the age of the wine.

Determination of Gelatine Index

The gelatine index was determined using the method proposed by Glories (1978).

The gelatine index estimates the percentage of tannins that precipitate in the presence of proteins (Glories, 1978). The sensory sensation known as astringency is as a result of tannins in the wine binding and precipitating the mucus proteins in the saliva. Gelatin is added to the wine to bind and precipitate the tannins in the wine.

The affinity to gelatine is visible from the decrease in absorbance at 280 nm, observable after the reaction course.

To 10 mL of sample 1 mL of water is added and the OD at 280 nm against water is measured, after dilution 1:50 (d_0). Simultaneously to other 10 mL of solution into a test tube 1 mL of gelatin solution is added, nitrogen is insufflated in the head space, and, once plugged, the test tube is placed at 10°C for 72 hours. After this period, the sample is centrifuged and the OD at 280 nm of the supernatant is measured, against water, after dilution 1:50 (d_{72}).

$$\text{Gelatine index} = \frac{d_0 - d_{72}}{d_0} \times 100$$

Determination of Catechins Content

The determination of flavanols in wine is carried out according to a method proposed by Zironi et al. (1992), using of 4-(dimethylamino)-cinnamaldehyde (DAC).

The assay is performed using a chromogen reagent of 1g DAC dissolved in 250 mL of 37% HCl and 750 mL methanol.

To 1 mL sample diluted 1/25 (v/v), there shall be added 5 ml of DAC solution, and then the maximum value of absorption at 640 nm against a blank prepared by substituting the sample with 1 mL 10% ethanol is measured. Catechins are expressed in mg/L, in reference to a calibration curve obtained with (+)-catechin in ethanol 10% (or multiplying by 30 and the number of dilutions).

Statistical Analysis

An analysis of variance (ANOVA) with 95% confidence level was carried out for each variable response on samples 1 and 2 in order to test the model significance and its suitability obtained using Modde 8.0.2 Umetrics software.

The results obtained from the tests replicated 3 times (sample 3), instead, have been analyzed with ANOVA through the Statistica for Windows 7.0.

RESULTS

Phenolic Compounds Extraction by Ultrasound

As described in our previous work (Ferraretto et al., 2013), we found that the 20 kHz ultrasound treatment in a solid-liquid medium, such as pressed crushed grapes, was responsible for accelerating the release and the dissolution of phenolic compounds contained in epidermal tissues.

The total polyphenol index, the anthocyanins content and the color intensity, related to the control sample extracted in a wine-like solution and in acidified methanol, proportionally increased (50% or more) with the time and amplitude of treatment for all grapevine varieties tested (analyzed) (Figure 4 and 5). Under the same treatment conditions, the total polyphenol index and the anthocyanins content proved different depending probably on the characteristics (cell wall maturity, pectin content) of the different grapes varieties (Ferraretto et al., 2013).

The tests carried out to evaluate the immediate effect on the color showed an important interaction between the cavitation phenomena and skin cell wall, leading to a higher color intensity and anthocyanins content. The compound extraction increases directly with the amplitude level ($r^2 = 0.9147$) (Figure 6).

Also the micro-vinification of samples treated with ultrasound provided interesting results, as reported by Ferraretto et al. (2013). In the study, the crushed treated vinified samples had higher polyphenol content in comparison with the reference at the beginning of maceration. The same was found at the end of the process and it influenced the total polyphenol index more than the anthocyanins content.

The different series gave different results, especially sample 2 with a higher initial phenolic content. Generally, it is possible to affirm that the variability of the ultrasound treatment effect depends on the quality of the raw material (Table 1).

Focusing on process optimization, a maximum of three days maceration time could be saved with respect to the five days in the classical vinification. The results have shown a reduction from 1 to 3 days of maceration time for both the phenolic and anthocyanins content. Consequently, the ultrasound treatment could be considered an auxiliary technology during maceration (Figure 7).

Table 1. Comparison of the total polyphenols index (TPI) and anthocyanins (ANTHO) on wine before (t0) and after ultrasound treatment, and after treatment 4 days of maceration (Merlot), with respect to the control (5 days of maceration) for the samples 1, 2 and 3, in the microvinification trials. US = sample treated with ultrasound; Control = reference sample

	Sample 1				Sample 2				Sample 3			
	US		Control (5 days)		US		Control (5 days)		US		Control (5 days)	
maceration time (days)	TPI	ANTHO (mg/L)	TPI	ANTHO (mg/L)	TPI	ANTHO (mg/L)	TPI	ANTHO (mg/L)	TPI	ANTHO (mg/L)	TPI	ANTHO (mg/L)
t0	7.5	17	7.5	17	19	258	19	258	12	25	12	25
After treatment	18	63	9	38	24	413	19	258	21	43	12	25
4 days	34	406	27	375	49	897	49	927	32	508	27	417

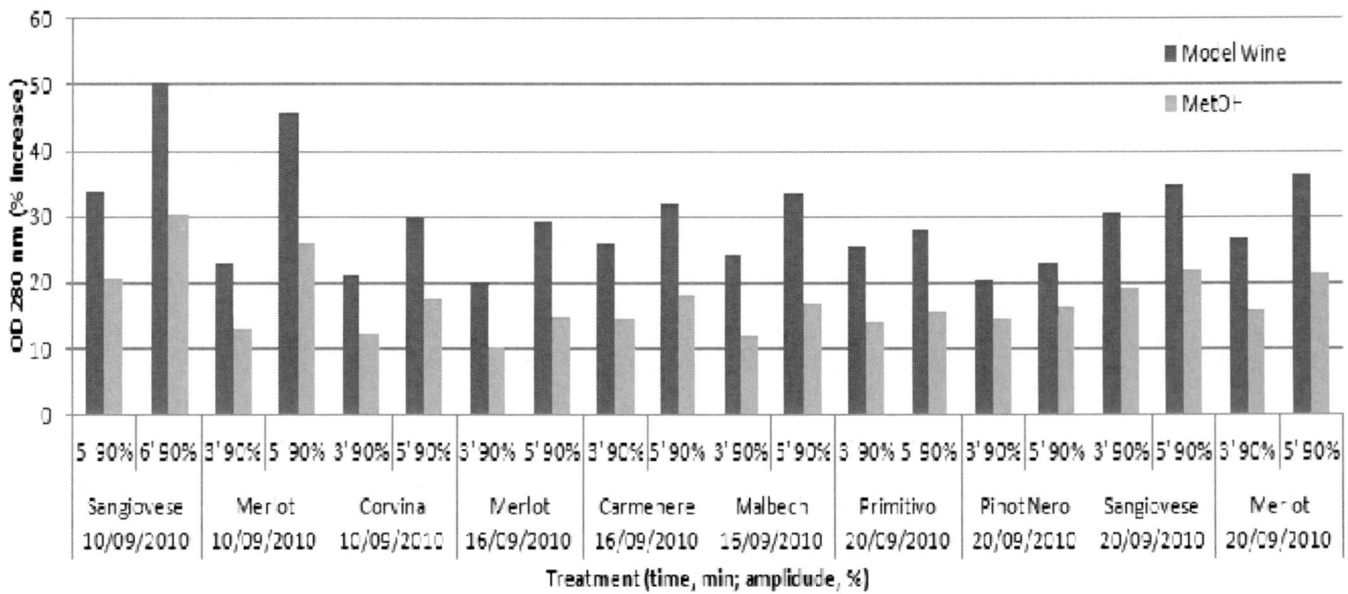

Figure 4. Percentage increase of OD 280nm of some red grapes varieties treated with ultrasound (3 and 5 minutes at 90% of amplitude) compared to the control extracted in model wine and in acidified methanol, for the phenolics extraction from the grape berry skin trials.

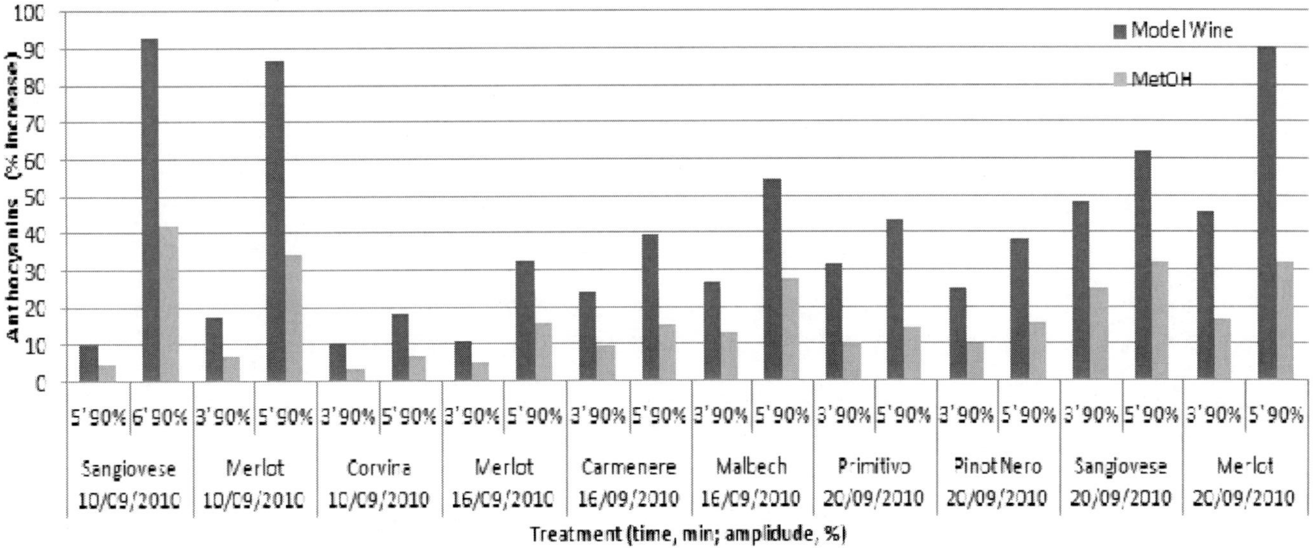

Figure 5. Percentage increase of anthocyanins content of some red grapes varieties treated with ultrasound (3 and 5 minutes at 90% of amplitude) compared to the control extracted in model wine and in acidified methanol, for the phenolics extraction from the grape berry skin trials.

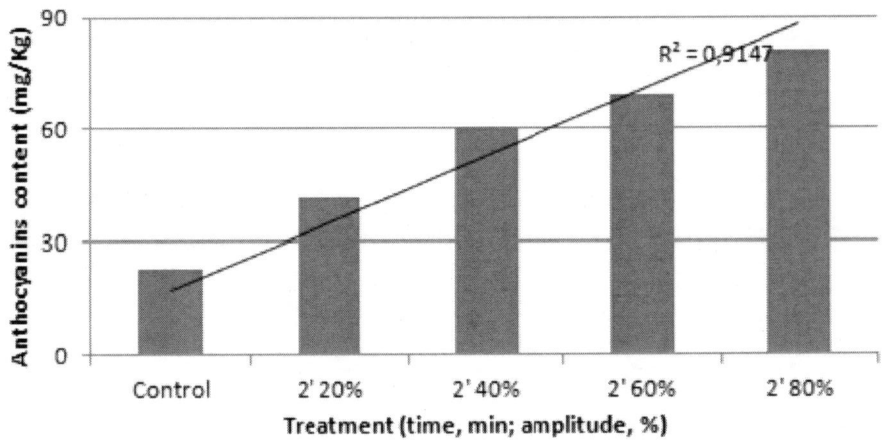

Figure 6. Change in anthocyanins content of the sample treated with ultrasound at different amplitudes (20, 40, 60, 80%) for the same time (2 minutes) for the phenolic extraction from the grape berry skin trials.

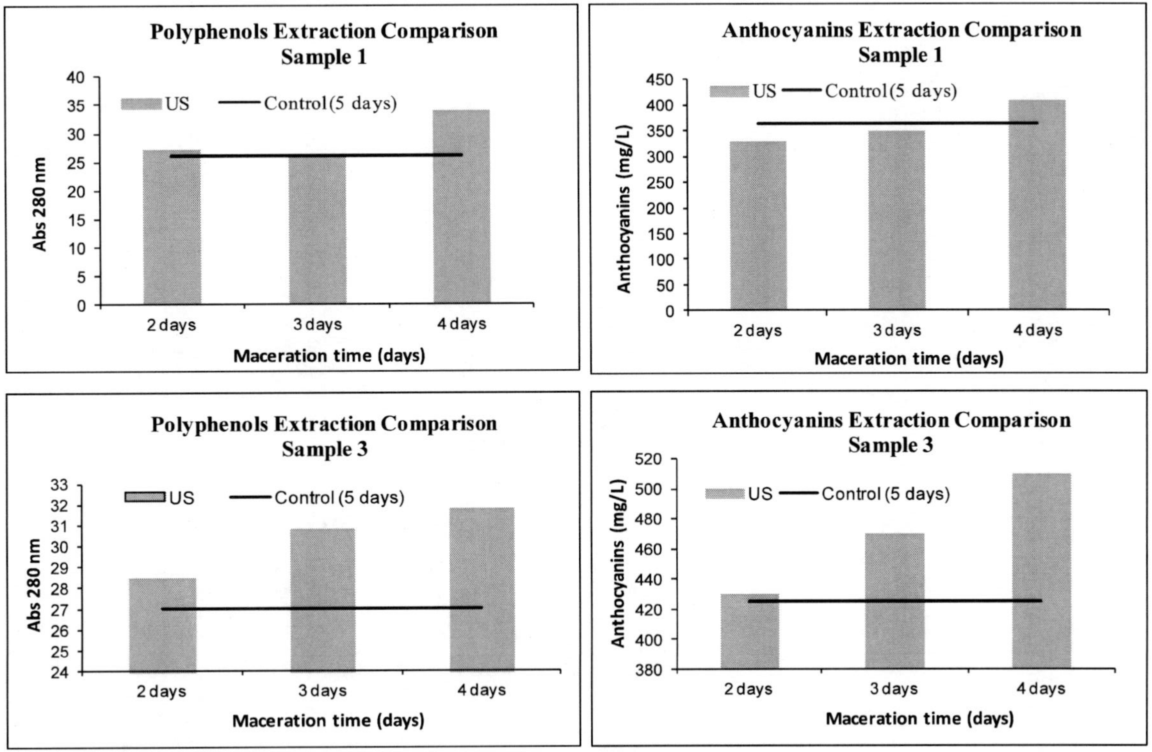

Figure 7. Comparison of the total polyphenols index (Abs 280 nm) and anthocyanins on wine after 2, 3 and 4 days of maceration (Merlot), with respect to the control (5 days of maceration) for the samples 1 and 3, in the microvinification trials. US = sample treated with ultrasound; Control = reference sample.

Effects on Young Red Wine

In our latest study (Ferraretto and Celotti, 2016), we found that the addition of mechanical and chemical energy to the young red wine causes some modifications: by comparing the three samples, we noticed that the ultrasound applied to wines with different composition leads to different results.

Since the temperature rises during the treatment, resulting into possible changes on the treatment effects and final wine quality, this parameter was monitored and recorded during the treatments. The increase in temperature from the standard 20°C is related to treatment times and power. The temperature increased to a maximum of 43.5°C.

This increase of temperature is not to be considered a problem in a potential future industrial application; the increase which could occur in few minutes of treatments does not cause a loss of wine color quality, as found by Galvin's studies (1992). Besides, it has to be considered that it is possible to carry out the treatment in controlled temperature conditions, if necessary, to avoid undesirable increases of wine temperature. The effects of the different treatments on color and predisposition to accelerated aging of the young red wines treated are reported in Table 2.

Table 2. Effect of ultrasound treatment on color and phenolic compounds of samples 1 and 2 (analyzed by experimental design; means ± standard deviation (n = 3) for the central point). au = absorbance unit

Treatment conditions	Sample 1			Sample 2		
	Color Intensity (au)	Anthocyanins (mg/L)	Catechins (mg/L)	Color Intensity (au)	Anthocyanins (mg/L)	Catechins (mg/L)
Tq	5.957	208	316	7.169	441	401
1' 30%	6.168	228	324	7.825	451	434
3' 30%	6.035	223	329	8.027	455	423
5' 30%	6.294	219	318	7.973	452	444
1' 60%	5.992	229	341	7.836	462	413
3' 60%	6.006±0.008	220±10.7	331±6.0	8.035±0.004	452±12.1	420±1.7
5' 60%	6.117	216	341	8.268	447	434
1' 90%	5.873	218	326	7.916	446	404
3' 90%	5.947	228	329	7.989	460	401
5' 90%	6.206	230	342	8.145	455	401

It is interesting to notice that no negative consequences on the anthocyanins, either loss of color, appeared after the treatment. Free anthocyanins were not modified, thus confirming their chemical stability under the treatment conditions applied. The stability of free anthocyanins is very important because they can combine later with tannins to obtain stable macromolecular complexes.

The most interesting results were achieved with tannic compounds. The catechin content analysis showed an increase in values on all samples; the treatment probably promoted the liberation of the monomeric catechins from tannins with different effects on their reactivity (tannins reactivity). Statistical parameters of the response surface model present a good correlation coefficient for catechins (r^2 = 0.8989, r^2 Adj. = 0.7977) for the sample 1. A significant increase of catechins was observed only for high values of amplitude and longer times of treatment. In the others conditions the results highlight the interaction between variables and accordingly the need of ultrasound treatments with specific combinations of amplitude and time based on the result expected. We could suppose different treatment conditions depending on the ratio between the polyphenols and mainly in function of the tannic structure of the wine, closely related to the content of catechins.

The HCl index showed different results on the samples. The ultrasound treatment which led to an increase or a decrease of these results could suggest the polymerization of the tannins (i.e., in sample 2 the values at 24 hours were always higher than the ones at 7 hours with a good correlation coefficient – r^2 = 0.8723, r^2 Adj. = 0.7447) or a breakage of the same tannins in the case of sample 3. It may depend on the different wine composition, including total content and of polyphenols and the ratio among different classes of molecules.

According to these results, we affirm that a few minutes treatment could be a valid approach to manage some parameters in the aging of red wines; in particular, the increase of HCl index related to tannins polymerization contributes to the wine astringency reduction, sensory feature required for red wines. According to El Darra et al. (2013) and Boulton (2001) the higher degree of anthocyanins-tannins copigmentation may conduct to enhance color stability. Considering the ultrasound treatment has a different effect on different wines, this approach could be useful, if confirmed by other experiments, to manage the tannin polymerization level to be able to control the chemical and physical stability of the tannic macromolecules and their astringency. A depolymerization could preserve tannins from colloidal precipitation risk, while their polymerization (an increase of the HCl index) would result in an astringency decrease. The index of ethanol also led to significant results, as its changes might have supported a rearrangement, probably due to an increasing reactivity of the polysaccharides and tannin compounds (Table 3).

The astringency, evaluated by gelatin index, of the sample 3 increases with short and mild treatment, while is reduced producing a pleasant sensation with stronger treatment conditions, particularly at the conditions of 5 minutes at 60% of amplitude and 1, 3 and 5 minutes at 90% of amplitude (Table 3). The different modifications of the gelatin index, directly related to astringency, might confirm the hypothesis that the ultrasound treatment conditions can be modulated depending on the tannic composition of the wine, making it possible to manage astringency perception in a short time. It could also be interesting to study the ultrasound effect on the interactions among tannins and polysaccharides, and the consequent changes of the organoleptic characteristics.

Table 3. Effect of ultrasound treatment on phenolic compounds of samples 1, 2 and 3 (analyzed by experimental design; means ± standard deviation (n = 3) for the central point); for sample 3 different letters represent significant differences according to ANOVA and Tukey HSD test ($p < 0.05$)

Treatment conditions	Sample 1			Sample 2			Sample 3		
	EtOH Index (%)	HCl Index 24h (%)	Gelatin Index (%)	EtOH Index (%)	HCl Index 24h (%)	Gelatin Index (%)	EtOH Index (%)	HCl Index 24h (%)	Gelatin Index (%)
Tq	20	38	24	8	24	27	17abcd	51±1d	48±0de
1' 30%	24	60	23	8	27	29	11a	53±3d	52±1e
3' 30%	12	49	18	11	31	32	12ab	48±3cd	52±3e
5' 30%	11	48	16	13	31	26	14abc	54±1d	53±2e
1' 60%	13	45	16	8	26	27	20bcd	48±5bcd	46±2cde
3' 60%	17±2.6	46±3.3	15±3.4	11±1.8	31±0.8	27±1.6	20bcd	47±2bcd	40±1bcd
5' 60%	20	42	24	25	33	26	22d	50±2d	39±2bc
1' 90%	12	38	25	11	33	28	17abcd	42±2abc	34±4ab
3' 90%	9	44	22	14	28	28	17abcd	41±1ab	29±6a
5' 90%	13	40	15	13	34	27	22cd	38±1a	39±4bc

CONCLUSION

The application of ultrasound in winemaking represents, first of all, a possibility to optimize fermentation technology and, therefore, a better management of red grape vinification. Furthermore, the application of ultrasound can be considered as a continuous pre-treatment of crushed red grapes before loading the vinification tank, in view of the very short treatment times.

The tests carried out on different grape varieties showed indeed how a few minutes of treatment at different amplitude levels can achieve an enhancement in the extraction of phenolic compounds, and in terms of time, a progressive reduction in the time of the classic maceration.

From the results obtained on wines after treatment with ultrasound, some indications suggest the use of the ultrasound to promote the polymerization of the phenolic compounds as the wine matures and therefore accelerate the aging process of wines. This technology can give the best result in the treatment of young, well colored wines featured by an immature tannic structure which is still evolving.

A brief ultrasound treatment could replace or integrate the traditional aging and stabilization techniques that are often constrained by the length of the kinetics of reactions between polyphenolic molecules.

It is also necessary to further investigate on the kinetics and response times to optimize the ultrasound technology on the treatment of red wines. Different results on different wines imply an ultrasound effect shown on disparate amounts of phenolic fractions. The results, based on variable treatment time and percentage of amplitude, have to be considered and evaluated to set the best treatment conditions for each wine. Moreover the treatment of white fermentation lees due to the yeast cell disorganization and the easier particulate cell leakage promotes the lysis of yeast with a rapid release of colloids, polysaccharides, and mannoproteins, and stabilize the wine. These effects can provide also a possible reduction in wine ageing on lees.

The results obtained confirm the extractive and lytic effects of the ultrasonic treatment on food matrices and could be applied certainly on a technological level to optimize some technological processes in the wine industry.

Ultrasound can be considered a new and sustainable technology for low impact winemaking: this technique could reduce the use of sulfur dioxide and enzyme for the extraction of phenolic compounds and, stabilize the wine without the use of exogenous adjuvants from yeast.

A winemaking with ultrasounds could result favorable to optimize costs of process, reduce time, and re-utilize by-products of vinification respecting the quality of the raw material.

REFERENCES

Bate-Smith, E-C. (1954). Leucoanthocyanins. I. Detection and identification of anthocyanin formed from leucoanthocyanins in plant tissues. *Biochemical Journal*, 11, 1153–1156.

Bates, D., Patist, A. (2010). "Industrial applications of high power ultrasonics in the food, beverage and wine industry." In *Case studies in novel food processing technologies: innovations in processing, packaging, and predictive modelling*. Cavitus Pty Ltd, 32 Spring Gully Rd, Crafers, SA 5052, Australia, 197, 119–138.

Boulton, R. (2001). The Copigmentation of Anthocyanins and Its Role in the Color of Red Wine: A Critical Review. *American Journal of Enology and Viticulture,* 52, 67–87.

Bussa-Valverde, N., Gómez-Plaza, E., López-Roca, J.M., Gil-Muñoz, R., Bautista-Ortín, A.B. (2011). The extraction of anthocyanins and proanthocyanidins from grapes to wine during fermentative maceration is affected by the enological technique. *Journal of Agricolture and Food Chemistry,* 59, 10, 5450–5455.

Cacciola, V., Ferran Batlló, I., Ferraretto, P., Vincenzi, S., Celotti, E. (2013). Study of the ultrasound effects on yeast lees lysis in winemaking. *European Food Research and Technology,* 236, 311-317.

Castellari, M., Matricardi, L., Arfelli, G., Amati, A. (2000). Level of single bioactive phenolics in red wine as a function of the oxygen supplied during storage. *Food Chemistry,* 69, 61–67.

Chandrapala, J., Oliver, C., Kentish, S., Ashokkumar, M. (2012). Ultrasonics in food processing. *Ultrasonics Sonochemistry*, 5, 975–983.

Chang, A.C., Chen, F.C. (2002). The application of 20 kHz ultrasonic waves to accelerate the aging of different wines. *Food Chemistry,* 79, 501–506.

Chang, A.C. (2004). The effects of different accelerating techniques on maize wine maturation. *Food Chemistry,* 86, 61–68.

Chang, A.C. (2005). Study of ultrasonic wave treatments for accelerating the aging process in a rice alcoholic beverage. *Food Chemistry,* 92, 337–342.

Chemat, F., Zill, E.H., Khan, M.K. (2011). Applications of ultrasound in food technology: processing, preservation and extraction. *Ultrasonics Sonochemistry*, 18, 813–835.

Clark J.P. (2008) An update on ultrasonics, Food Technology (26) 75-77.

Demirdöven, A., Baysal, T. (2009). The use of ultrasound and combined technologies in food preservation. *Food Reviews International*, 25, 1–11.

El Darra, N., Grimi, N., Maroun, R.G., Louka, N., Vorobiev, E. (2013). Pulsed electric field, ultrasound, and thermal pretreatments for better phenolic extraction during red fermentation, *European Food Research and Technology,* 236, 47–56.

Feng, H., Lee, H. (2011). “Effect of Power Ultrasound on Food Quality”. In *Ultrasound Technologies for Food and Bioprocessing*. Springer, London, 559–582.

Ferraretto, P., Cacciola, V., Ferran Batllò, I., Celotti, E. (2013). Ultrasound application in winemaking: grape maceration and yeast lysis. *Italian Journal of Food Science,* 25, 160–168.

Ferraretto, P., Celotti, E. (2016). Preliminary study of the effects of ultrasound on red wine polyphenols. *Cyta – Journal of Food,* in press.

Galvin, C. (1992). *Etude de certaines réactions de dégradation des anthocyanes et de leur condensation avec les flavanols; conséquences sur la couleur des vins*. Thèse de Doctorat. Université de Bordeaux II.

García Martín, J.F., Guillemet, L., Feng, C., Sun, D.W. (2013). Cell viability and proteins release during ultrasound-assisted yeast lysis of light lees in model wine. *Food Chemistry*, 141, 934–939.

García Martín, J.F., Sun, D-W. (2013). Ultrasound and electric fields as novel techniques for assisting the wine ageing process: The state-of-the-art research. *Trends in Food Science and Technology,* 33, 40–53.

Glories, Y. (1978). *Recherches sur la matière colorante des vins rouges.* Thèse de Doctorat. Université de Bordeaux II.

Glories, Y. (1984). La couler des vins rouges, 1e partie. Les equilibres des anthocyanes et tannins. *Connaissance de la vigne et du vin,* 18, 195–217.

Gong, C., Hart, D.P. (1998). Ultrasound induced cavitation and sonochemical yields. *Journal of the Acoustical Society of America*, 104, 2675–2682.

Guerrero, S., López-Malo, A., Alzamora, S.M. (2001). Effect of ultrasound on the survival of Saccharomyces cerevisiae: influence of temperature, pH and amplitude, *Innovative Food Science and Emerging Technologies*, 2, 31–39.

Jackson, R.S. (2008). *Wine Science: Principles and Applications.* San Diego, USA, Academic Press.

Jiranek, V., Grbin, P., Yap, A., Barnes, M., Bates, D. (2008) High power ultrasonics as a novel tool offering new opportunities for managing wine microbiology, *Biotechnology Letters*, 30, 1, 1–6.

Knorr, D., Zenker, M., Heinz, V., Lee, D.U. (2004). Applications and potential of ultrasonics in food processing. *Trends in Food Science and Technology*, 15, 261–266.

Lieu, L.N., Le, M.V.V. (2010). Application of ultrasound in grape mash treatment in juice processing. *Ultrasonics Sonochemistry*, 17, 1, 273–279.

Mason, T.J., Paniwnyk, L., Lorimer, J.P. (1996). The uses of ultrasound in food technology. *Ultrasonics Sonochemistry*, 3, 253–260.

Mason, T.J. (1998). "Power ultrasound in food processing-The way forward." In *Ultrasound in food processing*. Blackie Academic and Professional, London, 105–126.

Masuzawa, N., Ohdaira, E., Ide, M. (2000). Effects of ultrasonic irradiation on phenolic compounds in wine. *Japanse Journal of Applied Physics,* 39, 2978–2979.

McClements, D.J. (1995). Advances in the application of ultrasound in food analysis and processing. *Trends in Food Science and Technology*, 6, 293–299.

Povey, M.J.W., McClements, DJ. (1988). Ultrasonics in food engineering. Part I: introduction and experimental methods. *Journal of Food Engineering*, 8, 217–245.

Ribéreau-Gayon, P., Stonestreet, E. (1965). Le dosage des anthocyanes dans le vin rouge. *Bulletin de la Société Chimique de France*, 9, 2649–2652.

Ribéreau Gayon, P. (1970). Les dosage des composés phénoliques totaux dans les vins rouges. *Analytica Chimica Acta,* 52, 627–631.

Roberts, R.T. (1993). High intensity ultrasonics in food processing. *Chemistry and Industry*, 15, 119–121.

Sacchi, K.L., Bisson, L.F., Adams, D.O. (2005). A review of the effect of winemaking techniques on phenolic extraction in red wines. *American Journal of Enology and Viticolture*, 56, 197–206.

Spranger, M.I., Climaco, M.C., Sun, B.S., Eiriz, N., Fortunato, C., Nunes, A., Leandro, M.C., Avelar, M.L., Belchior, A.P. (2004). Differentiation of red winemaking technologies by phenolic and volatile composition. *Analytica Chimica Acta,* 513, 151–161.

Sudraud, P. (1958). Intérpretation des courbes d'absorption des vins rouges. *Annales de technologie agricole*, 7, 203-208.

Sun, B.S., Spanger, M.I., Roque-Do-Vale, F., Leandro, M.C., Belchior, A.P. (2001). Effect of different winemaking technology on phenolic composition in Tinta Miuda red wines. *Journal of Food Chemistry*, 49, 5809–5816.

Sun, D-W., Li, B. (2003). Microstructural change of potato tissues frozen by ultrasound-assisted immersion freezing. *Journal of Food Engineering,* 57, 337–345.

Suslick, K.S. (1989). The chemical effects of ultrasound. *Scientific American*, 260, 80–86.

Suslick, K.S. (1998). "Homogeneus sonochemistry". In *Ultrasound: Its Chemical, Physical, and Biological Effects*. VCH Publishers, New York, 123–163.

Tao, Y., Sun, D-W. (2015). Enhancement of food processes by ultrasound: a review. *Critical Reviews in Food Science and Nutrition*, 55, 570–594.

Yap, A., Jiranek, V., Grbin, P., Barnes, M., Bates, D. (2007). The application of high power electronics to enhance winemaking processes and wine quality. *Australian and New Zealand Wine Industry Journal*, 22, 2, 44–48.

Yap, A., Schmid, F., Jiranek, V., Grbin, P., Bates. D. (2008). Inactivation of Brettanomyces/Dekkera in wine barrels by high power ultrasounds. *Wine Industry Journal of Australia*, 23, 32–40.

Zhang, Q.A, Shen,Y., Fan, X.H., García Martín, J.F. (2016). Preliminary study of the effect of ultrasound on physicochemical properties of red wine. *CyTA-Journal of Food*, 14, 55–64.

Zheng, L.Y., Sun, D-W. (2006). Innovative applications of power ultrasound during food freezing process – a review. *Trends in Food Science and Technology*, 17, 16–23.

Zironi, R., Buiatti, S., Celotti, E. (1992). Evaluation of a new colourimetric method for the determination of catechins in musts and wines. *Wein-Wissenschaft*, 47, 1–7.

ABOUT THE EDITOR

Dr. Juan Francisco García Martín was born in Linares (Spain) in 1979. He earned a B.Sc. in Chemistry in 2001 and afterwards he obtained a Ph.D. summa cum laude (European Doctorate Mention) in 2007. Subsequently, he was postdoctoral researcher at the University of Granada, University College Dublin and the Spanish National Research Council. He was also lecturer at the Universities of Jaen and Granada, and currently he is lecturer at the University of Malaga.

INDEX

E

F

G

H

I

K

P

Q

R

S

T

U

V

W

X

Y